Challenging Students to
DO
Meaningful Mathematics

by
Joseph I. Stepans & Linda S. Hutchison

A resource for classroom teachers,
prospective teachers,
higher education faculty,
& staff developers

SP

SAIWOOD PUBLICATIONS
Montgomery, Alabama
ISBN 0-9649967-1-5

©1998

Challenging Students to DO
Meaningful Mathematics
by
Joseph I. Stepans & Linda S. Hutchison
University of Wyoming

Published and distributed by

SAIWOOD PUBLICATIONS
P. O. Box 242141
Montgomery, AL 36124

Telephone	(334) 277-3433
Toll-free	(800) 743-4787
FAX	(334) 277-0105
email	bsaigo@aol.com

Call, FAX, or email for ordering information.
Suggestions and corrections are welcomed!

ISBN 0-9649967-1-5

The activities, strategies, and models presented in this book are consistent with research on teaching and learning and the Standards that have been established by the National Council of Teachers of Mathematics.

Printing and binding by
WALKER PRINTING
Montgomery, Alabama

Printed on recycled paper

ABOUT THE AUTHORS

Joseph I. Stepans is currently a Professor of Science and Mathematics Education at the University of Wyoming. He has a B.S. in physics with a minor in mathematics, a Master's degree in physics, and a Ph.D. in science education. In addition to his experiences at the university level, he taught mathematics, physics, chemistry, physical science, and earth and space science in grades 8-12. Stepans has been active in researching and developing ways to deal with students' science and mathematics misconceptions since 1981, and has been published extensively. In addition to authoring and co-authoring books and numerous professional articles, he has shared his work at hundreds of presentations at state, national, and international conferences and workshops.

Linda S. Hutchison is an Assistant Professor of Mathematics Education at the University of Wyoming. She earned a Ph.D. in mathematics education from the University of Washington. She taught for several years in grades 4-12 mathematics classrooms in California. Her research interests include the preservice teacher's acquisition of pedagogical content knowledge, the preparation of mathematics teachers, and students' mathematical misconceptions. Dr. Hutchison has done presentations and workshops at local, state, and national levels.

ACKNOWLEDGMENTS

We are very grateful to:

John Russell of Natrona County Schools, Wyoming, for his helpful suggestions.

Barbara Woodworth Saigo for her many valuable contributions and efforts in editing, formatting, and illustrating the book.

Sandi Schlichting, of Idea Factory, Riverview, Florida, for her encouragement and assistance in initiating and developing this book.

Teachers, authors, and researchers whose works and ideas have been used in this book.

We give special thanks to:

Our **families** for their encouragement and support.

To the many **teachers and administrators** across the country who have been part of the workshops and conference presentations, and especially the Wyoming TRIAD (WyTRIAD) participants for their encouragement to produce this book.

Our **reviewers**, including:

Carol Jahn
Campbell County High School
Gillette, Wyoming

Gayle J. Jellum
Mathematics Department Chair
Twin Spruce Junior High School
Gillette, Wyoming

Suzanne Morrison, Director
University of Wyoming Lab School
Laramie, Wyoming

Diane Schmidt
Franklin Park Magnet School
Fort Myers, Florida

Dr. Morgan Simpson, Acting Dean
School of Education
Auburn University Montgomery
Montgomery, Alabama

Diana Wiig
Northpark Elementary School
Rock Springs, Wyoming

TABLE OF CONTENTS

NOTE: The remaining chapters are identified by numeral and topic. Each of the 16 topical chapters has the following organization:

>   A.   *Overview*
>   B.   *Background information for the teacher*
>   C.   *NCTM position*
>   D.   *Integrating problem solving, reasoning, communicating, and making connections*
>   E.   *Prerequisite skills*
>   F.   *Students' difficulties, confusion, and misconceptions*
>   G.   *Factors contributing to students' difficulties, confusion, and misconceptions*
>   H.   *Appropriate teaching strategies*
>   I.   *Teaching notes*
>   J.   *Materials*
>   K.   *Activities*
>   L.   *Assessment ideas*
>   M.   *Resources and references*

FOREWORD

According to primary teachers, when children first walk into the classroom they are excited and eager to learn. Teachers play a major role in sustaining the enthusiasm that children bring to school. It is sad, however, that many of these excited and eager children begin to dislike school, including mathematics, early in their schooling. As a result, many doors to the future will be closed to them, long before they are even aware of the opportunities they are forfeiting.

In America, mathematics has commonly been regarded as a difficult subject, which only a select group is expected to master and to benefit from. But lack of success in mathematics has broad implications, beyond the classroom. Mathematics has become an educational *filter*. If you do not do well in mathematics, you cannot enter certain courses, meaning that you cannot prepare successfully in many disciplines and, consequently, are blocked out of many professions.

Directions proposed by the National Council of Teachers of Mathematics should be implemented if our students are to succeed in a changing world. Our students must feel confident that they can *DO* mathematics. They must have the opportunity to make sense of mathematics and its applications. An understanding of mathematics should provide students opportunities and open doors, not the opposite.

Some students benefit from traditional instruction, which relies heavily on textbooks and written problems. They can grasp the abstractions and understand the algorithms. Most students, however, need to *DO* mathematics in order to make sense of it, to *UNDERSTAND*. Appropriate, alternative instructional strategies and activities are needed to reach these students, the majority.

National reforms acknowledge these concerns. *GOALS 2000* challenges us to help American students be first in mathematics and science in the world by the year 2000.

The *NCTM Standards* and documents such as *Everybody Counts* provide a vision for ALL students. **Their position is that mathematics must become a *pump* rather than a *filter* in the pipeline of American education.**

This book is one vehicle to help the classroom teacher make this vision a reality.
•   The approach to each topic in this book begins with the *learner,* deliberately identifying students' mathematical confusion, difficulties, and misconceptions *before* instruction on the topic.
•   By integrating problem solving, communicating, and reasoning, it attempts to help learners *meaningfully connect mathematics concepts to their lives.*
•   It suggests new roles for the teacher and the learner, making mathematics relevant and meaningful by encouraging the teacher to be a facilitator, actively involving the learner in *DOING* mathematics.

INTRODUCTION: LET'S ASK SOME QUESTIONS

How much do our students REALLY learn?

A group of students was given 5 questions and each student was individually interviewed. Below are the questions and students' sample responses.

1. **Students were told "Show the steps you would take to add 16 and 19."** Students showed their work as follows:

$$
\begin{array}{r}
① \downarrow \\
16 \\
+19 \\
\hline
35
\end{array}
$$

They said, "Add 9 and 6, and you get 15. Then you carry the 1, add 1, 1, and 1, and you get 35."

The students were asked, "Why did you carry the 1?" Approximately 71% of them gave responses like these: "I don't know." "It's just a formula." "The teachers always did that." "If you don't carry, you'll get a larger number." "I don't worry about why."

2. **Students were told "To compute 1/3 divided by 2/5, we write 1/3, change division to multiplication, and invert 2/5; in other words 1/3 x 5/2 = 5/6. Do you agree? Why do we do this?"**

All these students agreed with that process. When they were asked why it was done this way, 91% gave responses like these: " I don't know." "It's just a formula." "That's the way I was taught."

3. **Students were presented with the following examples. Two right triangles (a) are rearranged into two new configurations, b and c.**

Students were asked, "Which of the following statements is/are true and why do you think so?"

The areas of all three sets of triangles are the same.
Changing the arrangement of the triangles from <u>a to b</u> made the area bigger.
Changing the arrangement of the triangles from <u>a to c</u> made the area bigger.
Changing the arrangement of the triangles from <u>a to b</u> made the area smaller.
Changing the arrangement of the triangles from <u>a to c</u> made the area smaller.

Sample responses from 21% of these students included these: "One looks bigger." "The lengths are different." "The mass is the same." "You have affected the volume because of the formula for different shapes."

4. The students were asked, "Which is larger, 11/12 or 13/14, or are they equal?"

About 36% of the students said 11/12 was larger. These are sample responses: "11 is smaller than 13." " Fourteenths are smaller than twelfths." "11 wholes is greater than 14 wholes." " 12 into 11 gives you a larger number." "In real life 11 pieces of pizza cut into 12 is more than 13 pieces cut into 14. "

A few students said they were equal. One sample response was "They are equal because in both cases the numerator and the denominator differ by 1."

5. The students were asked "What is pi (π)?"

All the students gave the numerical value of either 3.14 or 22/7. None of the students knew that it was the ratio of the circumference to the diameter of a circle.

What is revealed in this study? These five concepts are *covered* in most elementary classes. Yet, whether they are *learned* in most elementary classrooms is less certain. In traditional instruction, the students merely need to come up with a correct answer. In this case, the interviewer posed different types of questions ~ questions that call for *understanding*.

What level were these students? THEY WERE COLLEGE SOPHOMORES.

Is the textbook way the only or best way to DO mathematics?

A group of teachers and their building administrators were presented with two containers, one labeled "ones" and the other labeled "tens." The "ones" container had single Popsicle sticks, while the "tens" container had bundles of ten sticks. The group was told that the rule is that the ones' container may not contain more than nine sticks. When it accumulates ten, the sticks should be bundled and transferred to the tens' container.

The participants were given the following task:

"You have two bundles of tens and seven ones placed in the appropriate containers. If you are given an additional three bundles of tens and five more ones, how will you go about combining (adding) them? What will be the result? You are to physically show what you would do."

There were various approaches. However, the majority of teachers began by combining the tens first, then the ones. When participants were asked to review the steps they took, many were surprised at the process they had used. Translating their actions into a mathematical algorithm, this is how it appeared:

	TENS	ONES
have	2	7
add	3	5
result	5	(12)

$$12 \rightarrow 10 \quad 2$$

$$10 \rightarrow \frac{5+1}{6} \qquad \frac{}{2}$$

In reflecting on their thinking, they were surprised that they intuitively combined the tens first and then the ones. They said that this was unlike the procedure they were used to ~ but IT WAS NATURAL! The procedure most textbooks have used traditionally is unlike what the majority of these participants chose. *Why do we continue imposing an unnatural algorithm on our students?*

Have the books really changed?

Nineteen graduate students completing their training in educational administration were given handouts that contained 6 sets of exercises for middle-level mathematics students. The participants were asked to work in their small groups of 3 or 4 and rate the mathematics problems and exercises on the handouts according to the following criteria:

- importance
- level of thinking targeted

- degree of relevance to the learner
- appropriateness for the students intended
- appeal to the learner
- consistency with research in mathematics education
- consistency with what they witness in a TYPICAL math class

After rating each set, the participants were asked to compare the odd-numbered and even-numbered problems and exercises on the given criteria.

The participants overwhelmingly rated the odd-numbered exercises as follows:

- more important
- required higher level thinking
- more relevant to middle-level students
- more appealing
- more consistent with the NCTM position

However, the participants felt that the even-numbered exercises were more consistent with what they typically see in a middle-level mathematics class.

The even-numbered problems and exercises were randomly selected from a textbook published in 1991. The odd-numbered problems and exercises came from an NCTM publication which was published in 1927.

Who is the audience for and what is the purpose of this book?

This book is intended for preservice and inservice early elementary through high school teachers.

It may be used as a resource by classroom teachers, prospective teachers, and higher education faculty in their content and methods classes. It is *not* a prescriptive book. It attempts to help preservice and inservice teachers of mathematics to create an environment in which students experience enthusiasm, thinking, and a willingness to DO mathematics.

What makes this book useful?

The content and approach of this book are designed to provide practical, useful assistance to teachers. It's particular features are:

- **It uses appropriate teaching strategies for each math concept.**

- **It identifies students' mathematical difficulties, confusions, and misconceptions, as well as some of the factors which are known to contribute** to these.

- It makes mathematics relevant and real for students.

- It meaningfully integrates problem solving, reasoning, communicating, and making connections with real situations illustrating each concept.

- It suggests new roles for the teacher and the students.

What are the issues?

To improve mathematics education for our students, many issues must be considered, especially new societal expectations.

1.   Explicit problems with mathematics education need to be identified. These problems include the following:

- low learning expectations for students
- lack of relevancy of materials
- inappropriate teaching strategies and materials
- mismatch of concepts with a student's developmental level
- lack of alignment of expectations, teaching, and assessment
- lack of consistency with what is known about learning and effective teaching
- stagnation of what is taught, how it is taught, and assessment used
- year-to-year redundancy of math topics addressed
- high-level mathematics for a small number of students
- mathematics used as a filter
- mathematics expected to be difficult
- perception that knowing the algorithm is the same as knowing the concept

2.   Changes in the mathematics curriculum should match changes in the world. By that we mean the world is different today than it was 30 years ago. What was considered "basic" about mathematical skills is no longer sufficient. *Basic now includes the ability to reason, to solve problems, to apply mathematics, and to use technology.*

3.   It is our responsibility to meet the varied needs of ALL students.

4.   A definition of mathematics needs to include <u>DOING</u> and <u>THINKING</u>.

5.   Teachers need to consider student readiness and understanding in planning teaching strategies and selecting learning materials.

6.   Many factors contribute to students' mathematics illiteracy and their anxiety toward mathematics. Included among these factors are the following:

- inappropriate concepts
- inappropriate sequence of concepts presented

- lack of attention to prerequisites
- inappropriate assumptions made by textbook companies and other curriculum developers
- year-to-year redundancy of math topics
- societal acceptance for not doing well, including the assumption that high math requires special skills which most students do not have
- inappropriate teaching strategies
- inappropriate assessments

7.   There must be alignment of NCTM recommendations, expectations of what students should learn, experiences to help them reach those expectations, and assessment of both the students' performance in regard to the expectations and the effectiveness of the instructional experiences.

8.   There are new roles for the learner, the teacher, and other education partners.

9.   Nearly all jobs require a knowledge of mathematics.

What are the results of inattention to these concerns? The impact of such practices may be reflected in our students' performance on the Third International Mathematics and Science Study (National Research Council, 1996).

Teaching strategies used in this book

It is our belief that to help students learn to DO mathematics, they have to be involved in *DOING* mathematics, not just repetitively practicing algorithms. To become effective *problem solvers*, the students have to be involved in *problem solving*. To become a mathematical *communicator*, the student has to be involved in *communicating* in mathematics and about mathematics. To learn to effectively search for and establish *patterns*, they have to be involved in *searching for and establishing* patterns. To become an effective *reasoner*, the student must be involved in *reasoning* with mathematics.

Much is now known about learning, and the research is advancing rapidly. In this book we will use what we know about learning to make mathematics and the learning of mathematics meaningful through a repertoire of strategies that actively engage students and create understanding.

Different concepts and skills call for different teaching strategies. Strategies targeted here include conceptual change, collaborative learning, concept mapping, discrepant events, mental model building, and discussion and demonstration.

Here are brief descriptions of those strategies:

1.   Conceptual change strategy

Based on a constructivist philosophy of learning, teaching for conceptual change is a strategy in which the teacher begins with what the individual student brings to the classroom. The student's ideas and views guide the learning process. The focus in this strategy is on creating meaningful change in the naive or incomplete ideas that the student brings to class. This strategy effectively blends hands-on activities, discussion, questioning, and collaboration to target a variety of learning modes.

2.   Collaborative learning

We prefer the broader concept of collaborative learning to formal cooperative learning. This is not just a semantic distinction, but recognizes that collaboration takes many forms. (One such form is cooperative learning.) Collaboration among students while they are DOING mathematics, as they are tackling problems which involve mathematics, incorporates many aspects of self-learning and peer-assisted learning. This strategy is particularly useful when a group of students uses each other's expertise and resources to find solutions to problems.

3.   Concept mapping

Designed initially by Novak and Gowin (1987) and used extensively in science education, concept mapping is a schematic device for presenting a set of concept meanings in a framework of propositions (visual statements). The mapping is a visual road map, showing some of the pathways one may take to connect meanings of concepts. Before using concept mapping with a mathematics concept, the teacher may go through an exercise with students on a familiar concept. An example may stem from students' everyday situations, such as a sport, food, animal, color, or anything about which the student can express related ideas.

4.   Discrepant events

A discrepant event is one in which something does not quite fit with one's current conception or expectation based on that conception. To appropriately use discrepant events, the teacher has to understand the algorithm the student is using and provide counter examples that make it obvious to the student that his/her approach does not work. The examples must be compelling and relevant to the learner. Confronting and working to explain discrepant events provides an excellent way for students to create their own understanding. Imagine that your students believe that when adding 27 and 34, you start adding from the left and get 27 + 34 = 511. When a teacher provides students with a concrete situation, using manipulatives or other means, students are stimulated to confront and correct the flaw in their methodology.

5.   Mental model building

Having students build their own mental models encourages them to overtly articulate and evaluate their ideas by comparing their mental models with those of other members of the class and outside sources, assessing how well the models explain observable phenomena. This approach also encourages students to question and judge the models proposed by experts.

Mental model building is effective in developing awareness of the need for models. It is useful in helping students refine existing models and build their own.

6. Discussion & demonstration

With this strategy, we recommend a dialogue between teacher and students and among students. The discussion should be student-centered, be encouraging, and provide the learner with the opportunity to communicate mathematical ideas. We agree that there are times when the teacher needs to give information, but in order to make mathematics meaningful, we suggest that the emphasis should be placed on *learning* rather than on a model of teaching that is mainly disseminating information.

What topics are addressed in this book and how are they organized?

Major topics that are used as chapter organizers for this book are:

1. Prenumber concepts
2. Operations
3. Logic
4. Patterns and relationships
5. Number theory
6. Two-dimensional geometry
7. Fractions and decimals
8. Three-dimensional geometry
9. Measurement
10. Ratio and proportion
11. Discrete mathematics
12. Spatial thinking
13. Algebra
14. Data and statistics
15. Probability
16. Trigonometry

Each chapter will be formatted in the following sections:

A. Overview of concepts to be included
B. Background information about the concepts for the teacher
C. NCTM position on the concept (areas to emphasize and deemphasize)
D. How to meaningfully integrate problem solving, reasoning, communication, and making connections
E. Prerequisite skills and knowledge
F. Students' difficulties, confusion, and misconceptions related to the

topic
G. Factors contributing to students' confusion, difficulties, and misconceptions
H. Appropriate teaching strategies
I. Teaching notes
J. Materials needed
K. Activities
L. Assessment ideas
M. Resources and references

Resources and references

Center for Occupational Research and Development. (1988). *Applied Mathematics*. Waco, TX: Author.

Changes in Mathematics. (1994). *Mathematics Leads the Way - Annenberg*. Washington, DC: Author.

Elliott, P. C., & Kenney, M. J. (Eds.). (1996). *Communication in mathematics K-12 and beyond: National Council of Teachers of Mathematics 1996 Yearbook*. Reston, VA: NCTM.

House, P. A., & Coxford, A. (Eds.). (1995). *Connecting mathematics across the curriculum: National Council of Teachers of Mathematics 1995 Yearbook*. Reston, VA: NCTM.

National Council of Teachers of Mathematics. (1989). *Curriculum and Evaluation Standards for School Mathematics*. Reston VA: Author.

National Research Council. (1989). *Every Body Counts-A Report to the Nation on the Future of Mathematics Education*. Washington DC: Author.

National Research Council. (1996). *Mathematics and Science Education: What Can We Learn from the Survey of Mathematics and Science Opportunities (SMSO) and the Third International Mathematics and Science Study (TIMSS)?* Washington DC: Author.

National Research Council. (1997). *Third International Measure of Mathematics and Science: A Report of the Center for Science, Mathematics and Engineering Education*. Washington DC: Author.

Novak, J. D., & Gowin, D. B. (1984). *Learning How to Learn*. New York: Cambridge University Press.

Reys, R. E., Suydam, M. N. , & Lindquist, M. M. (1989). *Helping Children Learn Mathematics*. Englewood Cliffs, NJ: Prentice Hall.

Schmittau, J. (1993). Vygotskian scientific concepts: Implications for mathematics education. *Focus on Learning: Problems in Mathematics, 15(2 & 3).*

Stepans, J. I. (1996). *Targeting Students' Science Misconceptions: Physical Science activities Using the Conceptual Change Model.* Riverview, FL: Idea Factory.

Stepans, J. I., Saigo, B. W., & Ebert, C. (1995). *Changing the Classroom from Within.* Montgomery, AL: Saiwood Publications.

Tobias, S. (1987). *Success with Mathematics.* New York: College Entrance Examination Board.

Webb, N. L., & Coxford, A. F. (Eds.). (1993). *Assessment in the Mathematics Classroom.* Reston, VA: NCTM.

1
PRENUMBER CONCEPTS

A. Overview

Concepts included in this chapter are: classifying, patterns, comparing (as many as, less than, more than), conserving, ordering, early measurements, spatial relations, and early number sense.

B. Background information for the teacher

For adults the symbol "3" means three of something. The reason that adults accept this has to do with years of experience associating these symbols with objects. This connection continues for adults as they operate on numbers. When they add, subtract, multiply, or divide, they use symbols, numerals that represent objects. Sometimes teachers move too quickly to numbers, without understanding what they may mean to children. A number like 15 may not necessarily mean one 10 and five 1's. For children this may mean 6: five 1's and one 1.

Even very young children feel the need to use quantitative ideas. When a child says, "I have more candy than you have." or "My brother is younger than I am." or "My sister has fewer marbles than I have." that child is using mathematics.

Before we start with numbers and operations on numbers with young children, it is important to provide them with opportunities to experience prenumber concepts. Prenumber experiences need to provide, in an integrated and inseparable way, confidence and physical, social, and intellectual development. Young children rely on the concrete, sensory, or pictorial. The child needs to manipulate concrete objects and gradually, through these experiences, form mathematical ideas and develop the ability to apply symbols. These experiences involve comparing some sets, using "as many as," "more than," "less than," and ordering sets.

After children have had the opportunity to deal with prenumber concepts, they will be ready to associate objects with numbers. Counting then takes meaning. These prenumber and counting experiences are prerequisites to making sense of operations on numbers such as addition, subtraction, multiplication, and division.

To prepare students for numbers and operations on numbers, we need to help them develop prenumber concepts. For number concepts to become meaningful, many processes and skills are needed. Different children will use different approaches as they are learning numbers and operations. The more varied and different situations that are provided for children, the better the chances that they will come to make sense of the abstraction of numbers and operations on numbers.

Mathematics educators have identified several skills as prerequisites to learning mathematics, particularly the mathematics associated with numbers, operations, shapes and their properties. These skills include the following:

1. Classifying

The ability to separate and distinguish one object from another is an important prerequisite and provides children the opportunity to begin using numbers naturally. The ability to sort and classify is a fundamental thinking skill. Children come to school already familiar with some classification, such as boys and girls, adults and children, hot and cold. They usually also have experiences such as classifying by size, color, and physical make up. As children continue with school, these classification and sorting skills become more accurate and more involved. Children move from looking at only one attribute to considering various and more complicated ones. Learning the skill of classification helps children not only in mathematics, but in other disciplines like reading, science, and music.

2. Patterns

Searching for and establishing patterns are important thinking skills for children to develop. In the patterns and relationships units, we will build on these skills. Patterns may be represented by concrete objects, pictures, events, symbols, or words. Patterning activities with words, pictures, sounds, and physical materials are effective stepping stones to discerning patterns they will encounter later.

3. Comparing, conserving, ordering, and measuring

Such skills as identifying "as many as," "less than," and "more than" are other important prerequisites to counting and operations on numbers.

Piaget observed that rarely can children under five conserve. Children may show us that they can count but not be able to conserve. Conserving numbers, length, amount, and area are important prerequisites to the learning of many mathematical concepts.

Putting sets in increasing then decreasing order are important skills which the children will encounter later. Ordering activities must specify an arrangement and a rule, such as largest to smallest, oldest to youngest, longest to shortest, or vice versa.

Having children use their own units of measurement is an important beginning. Using their hands, feet, pop cans, or Popsicle sticks to measure things is a good start. Measuring and ordering can go well together.

4. Spatial relations

Reading a map, doing an art project, and understanding science concepts are among the many areas requiring that students have spatial skills. We assume (sometimes without checking) that children understand attributes like front, back, up, down, on top, under, next to, reverse, far, and near. Many mathematics concepts build on the knowledge of these attributes.

5. Early number experiences (associating numerals with objects)

Estimating may provide a good beginning for early number experiences. Before *counting* the number of objects, have children *guess* the number of objects.

Before coming to school, children usually have been exposed to counting. Counting is the process of a child beginning with one, pointing to the object, and going on in sequence. Children count, corresponding the number to the object, until all the objects are matched with numbers. Other kinds of counting include skip counting; *e.g.*, counting by two, three, ten, etc., or counting every other or every third object.

C. NCTM position

Manipulatives and other physical models should be used to help children relate processes to their conceptual understandings and give them concrete objects to talk about.

In all situations, work with number symbols should be meaningfully linked to concrete materials.

Students will develop counting skills, which are essential for ordering and comparing numbers.

D. Integrating problem solving, reasoning, communicating, and making connections

There are many opportunities for students to become involved in problem solving, reasoning, using various methods of communicating their ideas and making connections between the classroom and outside-world experiences.

E.   Prerequisite skills

Since a child's mathematical life *begins* with prenumber concepts, prerequisite skills are not necessary.

F.   Students' difficulties, confusion, and misconceptions

Many young children have misconceptions related to classifying, ordering, conserving numbers, length, material, area, and volume.

G.   Factors contributing to students' difficulties, confusion, and misconceptions

We should not expect young children to have been born with the skills of classifying, measuring, ordering, patterns, comparing, and conserving. Maturity and experiences are needed for children to develop these skills.

H.   Appropriate teaching strategies

All of the activities in this chapter call for manipulatives. Two teaching strategies are targeted. Teaching for conceptual change is used in the activities "Conserving relations," "Measuring," "Frames of reference," and "Looking in the mirror." Discussion and demonstration are prevalent in the activities "Classifying using different attributes," "Comparing sets," "Ordering," "Spatial relations," "Patterns," and "Early number sense."

I.   Teaching notes

Gelman and Gallistel (1978) suggest that children who are ready to count need a sequenced set of teacher-led activities. These authors outline five principles which apply to determining the size of a set, as follows:

1. One-to-one correspondence involves ticking off the items in a collection, in which only one member of the collection is used as it is counted.

2. Counting must be arranged in a sequence of words that does not change.

3. Using cardinal numbers in a set tells how many.

4. Any collection of real or imaginary objects can be counted.

5. The order in which objects are counted is irrelevant.

J. Materials

blocks of different shapes and colors
sets of pattern blocks
mirror
bottles or jars (see Activity 9)

K. Activities

See following pages.

1. Classifying using different attributes

You are presented with the following group of items: several different kinds of fruit, pen, pencil, paper, book, box, cup, glass, bottle, spoon, fork, knife, doll, money.

Using one characteristic, put the objects into three groups.

Share your grouping with someone else.

With your partner, see if you can come up with another way of grouping the items.

How many of these things are edible?

How many are made of paper?

Working with your partner, group the objects and challenge another group to figure out how you grouped them.

2. Comparing sets

Provide children the opportunity to experience situations similar to those shown below. Challenge the children to say <u>why</u> they responded as they did to each situation.

  a. Are there as many, more, or fewer

 than **?**

 **c. Are there as many, more, or fewer**

 than **?**

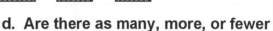 **d. Are there as many, more, or fewer**

 than **?**

 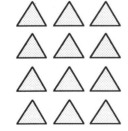 **f. Are there as many, more, or fewer**

 than **?**

3. Conserving relations

We are given 6 blocks placed in a straight line.

If we then place them in a circle do we still have the same number?

> Do we have more?

> Do we have less?

> Why do you think so?

We have a piece of string. If we make it into a square, a rectangle, or a circle, what will happen to the distance around?

> Will we have the same distance with each figure?

> Will we have less distance?

> Will we have more distance around?

> Why do you think so?

> How would you check your answer?

4. Ordering

Shown a set of blocks, put the smaller ones on this side. Put the larger ones on this side.

Given a set of pictures, find the youngest subject. Find the oldest subject.

Given a set of objects, find the largest. Find the next larger one, the next, and the next.

Given several objects, put them in order, and say why you put them in this order.

5. Spatial relations

Place the red block <u>next to</u> the green one

Place the yellow ball <u>on top of</u> the book

What is <u>between</u> the orange block and the green block?

Who is <u>closest</u> to the door?

Place the book <u>under</u> the chair.

Who is <u>next to</u> the window?

Stand <u>in front of</u> the door.

Given the chain and the blocks, set the green block <u>inside</u> the chain.

6. Measuring

Estimate: How many of these blocks tall (high) is this box?

Test your idea by measuring the box's height with the blocks.

Can you guess, how many blocks tall is the friend on your right?

How would you check your guess?

Do it!

Find other objects in the classroom to use as measuring tools, then measure books, desks, boxes and other items using those tools.

How many marbles will it take to fill this coffee can?

How many larger marbles would it take to fill the can?

7. Patterns

Using pattern blocks or Unifix cubes, complete the pattern that is started.

Start a pattern, then see if your friends can figure out the pattern and complete it.

8. Early number sense

Use different shapes to represent one object, two objects, three objects, four objects, and five objects.

Given numerals and the number of objects they represent, separate them and later match them.

9. Frame of reference

We have a container of water.

Draw what you think the water would look like if we were to tip the container to the left.

Draw what you think the water would look like if we were to tip the container to the right.

Draw what you think the water would look like if we were to lay the container on its side.

Get the necessary materials and test your predictions.

10. Looking in the mirror

Predict what your image would look like in the mirror if you raised your right hand.

Predict what your image would look like in the mirror if you raised your left hand.

Draw a picture of what this triangle would look like in the mirror.

Draw a picture of what the letter " E" would like in the mirror.

Test your ideas.

L. Assessment ideas

Observing students' performance is an effective way to keep track of children's actions and changes in their skills and conceptions.

Interviewing students on their understanding of prenumber concepts is also an effective assessment technique.

Observing inter-student communication during activities is very useful in monitoring their understanding.

M. Resources and references

Bruner, J. (1965). The growth of mind. *American Psychologist, 20,* 1007-1017.

Copeland, R. W. (1974). *Diagnostic and Learning Activities in Mathematics for Children.* New York: MacMillan.

Gelman, R., & Gallistel, C. R. (1978). *The Child's Understanding of Number.* Cambridge, MA: Harvard University Press.

Kennedy, L. M., & Tipps, S. (1994). *Guiding Children's Learning of Mathematics.* Belmont, CA: Wadsworth.

Labinowicz, E. (1980). *The Piaget Primer: Thinking, Learning, Teaching.* Menlo Park, CA: Addison-Wesley.

Labinowicz, E. (1985). *Learning from Children: New Beginnings for Teaching Numerical Thinking.* Menlo Park, CA: Addison-Wesley.

National Council of Teachers of Mathematics. (1989). *Curriculum and Evaluation Standards for School Mathematics.* Reston VA: Author.

Payne, J. N. (Ed.). (1975). *Mathematics Learning In Early Childhood.* Reston, VA: NCTM.

Reys, R. E., Suydam, M. N., & Lindquist, M. M. (1989). *Helping Children Learn Mathematics.* Englewood Cliffs, NJ: Prentice-Hall.

Tollefsrud-Anderson, L. (1993). *Counting and Number Conservation.* Paper presented at the Meeting of the Society for Research in Child Development. New Orleans, LA.

Troutman, A. P., & Lichtenberg, B. K. (1995). *MATHEMATICS- A Good Beginning.* Pacific Grove, CA: Brooks/Cole.

2
OPERATIONS

A. Overview

Concepts included in this chapter are: place value, operation algorithms, addition, subtraction, multiplication, division, and combining algorithms and situations.

B. Background information for the teacher

Place value

Place value is one of the most important features of the operation algorithms (addition, subtraction, multiplication, division). Students must be able to explain the logic behind the positions of digits (0 to 9) in a number so they can perform operations on the numbers.

Students can invent their own ways of calculating to solve problems--we want them to! These *invented algorithms* are procedures that students understand and are able to use and explain. They are often different from adults' algorithms. The use of invented algorithms to strengthen place value concepts is important (Kamii, Lewis, and Livingston, 1993).

Multiunit conception of numbers

Students also need to construct a *multiunit* conception of number (Fuson, 1990). In English the number words for hundred and thousand are regular. We say one hundred, two hundreds, and so forth. However the terms for ten are irregular in English, -teen or -ty depending on the number, like nineteen or twenty. Eleven and twelve are also irregular. This irregularity leads to one of the problems that children exhibit when they say words like "twentyteen." In some languages these numbers have words in the language that imply the place value. Research has suggested that the current textbook practice of doing operations on multidigit numbers before learning the sums up to 18 should be changed (Fuson, 1990; Baroody, 1990).

Addition

The situations that cause addition are the *joining* of two discrete sets to create a third set. This third set is the sum of the two smaller sets. The students' own terminology should be used in discussing this with them. Terms such as <u>add</u>, <u>put together</u>, <u>combine</u>, <u>join</u>, and <u>plus</u> could be used to develop understanding.

For example: Susie has 5 pens and Moira has 2 pens. How many do they have altogether?

Subtraction

The situations for subtraction include comparison, completion, part-whole, and take-away, as shown in the table below. In *comparison subtraction*, two sets of items are compared. For *completion subtraction*, a set is completed. For *part-whole subtraction*, a subset is determined by knowing the remainder of the set. The items in *take-away subtraction* are removed.

TYPE OF SUBTRACTION	EXAMPLE
Comparison	Taneka has 5 pens and Sally has 2 pens. How many more pens does Taneka have?
Completion	John has 3 tokens. If he needs 5 tokens to ride the Ferris Wheel, how many more tokens does he need?
Part-whole	There are 15 children in the class. If 8 are girls, how many of the children in the class are boys?
Take-away	Jose had 6 marbles. He gave 2 to Robert. How many marbles does Jose have left?

Multiplication

Multiplication has three models that students need to understand: repeated addition, array, and Cartesian product. To understand these models requires students to conceive of multiplication in different ways, as shown in the table below.

TYPE OF MULTIPLICATION	EXAMPLE
Repeated addition	Three students each brought 2 toys from home How many toys were brought from home?
Array	A large egg carton can hold 3 rows of eggs with 6 eggs in each row. How many eggs can it hold?
Cartesian product	The local ice cream store has a choice of cones: sugar and cake. If there are 5 flavors today, how many different kinds of single-scoop ice-cream cones could you create?

The *repeated addition model* features the same number being repeatedly added, such as finding how many Popsicle sticks are in four groups of ten. The groups could be counted as ten, twenty, thirty, forty Popsicle sticks, as shown below, and be written as $10 + 10 + 10 + 10 = 40$.

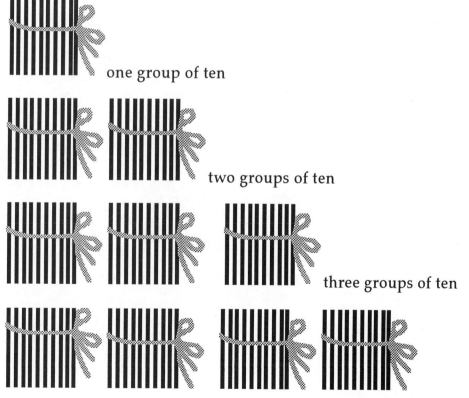

one group of ten

two groups of ten

three groups of ten

four groups of ten = forty sticks

Knowledge of *skip counting* (counting by twos, threes, fives, etc.) is useful for the repeated addition model.

The *array model* of multiplication is one in which thinking about the situation implies a picture that is rectangular, such as a grove, a parking lot, cans in a case, graph paper, students in rows in a classroom, or any objects that are commonly arranged in rows and columns.

For example, how many eggs are in an egg carton that has two rows and six columns?

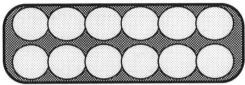

The *Cartesian product model* is appropriate for problems such as this classic situation: "I have four shirts and three pairs of pants. How many different outfits (combinations) can I make with these shirts and pants?" This type of situation requires the matching of one object to another and the answer reflects all of the combinations possible. It is much like a Cartesian graph that has ordered pairs (x, y) to represent points (or single matches) that can satisfy the problem.

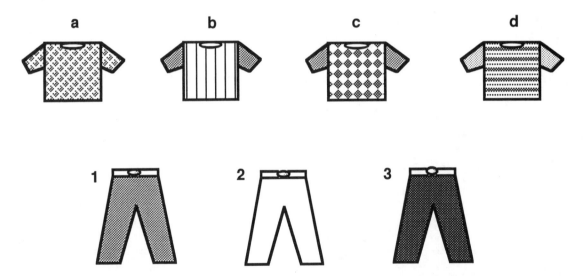

The choices are (a,1), (b, 1), (c, 1), (d, 1), (a,2), (b, 2), (c, 2), (d, 2), (a,3), (b, 3), (c, 3), and (d,3). There are 12 different possibilities.

Division

Two conceptions of division are important for a thorough understanding: partitive division and measurement division, as shown below.

Partitive division: $\dfrac{\text{whole}}{\text{number of groups}} = \text{number in each group}$

Measurement division: $\dfrac{\text{whole}}{\text{number in each group}} = \text{number of groups}$

Partitive division is the situation where a group of objects is *shared equally* among a group of people or other objects.

For instance, if Kim has 20 candies and 3 friends, how many candies would Kim and each friend get if they shared them equally? Kim could solve this by passing out each candy, one by one in turn, until none remained and everyone had an equal amount, as shown in the following figure.

For step one, Kim would distribute one candy to each person.
For succeeding steps, Kim would continue to parcel out candies, one at a time, until everyone had two, then three, then four, then five.

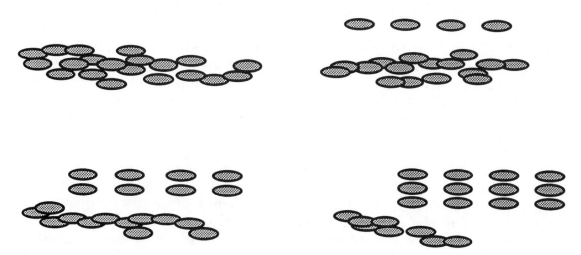

Kim would continue to give out candies until everyone had an equal amount and none remained. Each of the four children would get five pieces of candy, or five pieces per set.

If fewer remained than the number of sets, then the remainder would have to be split evenly, if possible. This splitting could result in a fraction of a piece of candy going to each child. If, however, it is not possible to split the leftover candy equally, it may be left as a remainder.

In some cases, like children getting on a bus for a field trip, it could be that an additional bus is needed for the remainder so the answer could be expressed as a whole number.

In *measurement division* equal parts are removed from the whole and the question is asked, "How many parts can you get out of the whole?" If a costumer designer has twenty yards of material and needs to make capes, using 5 yards for each cape, how many capes could be made? The twenty yards of material would be divided into five-yard increments and the answer would tell you the number of capes possible, in this case four. In measurement division, the number in each set is fixed and the question is really how many complete sets you can have. This is equivalent to repeated subtraction.

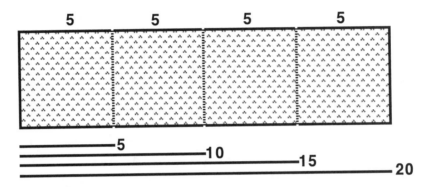

How should algorithms and operational situations be connected?

Algorithms should be learned in a variety of contextual situations to help students develop a better understanding. Students often perceive an operation in only one way. For instance, "take-away" is a common perception of subtraction. Many students do not understand the part/whole, comparison, and completion subtraction situations, and when they encounter them they are likely to add rather than to subtract.

The teacher encourages appropriate concept development by explicitly taking into account the diverse kinds of situations mentioned here and providing children a variety of concrete problems. Manipulatives and word problems emphasize how the situations fit with the symbolic algorithms. Also, word problems involving more than one operation force children to think more deeply about what is going on in the problem.

Concrete experiences should also be connected to the symbolic algorithms. For example, when adding 56 + 27 (see following illustration), students should be helped to connect the symbolic representation of the numerals to manipulatives (power of ten blocks) and the process of *regrouping*. As noted in the introduction, it is okay for children to add the tens first and then regroup the ones. This natural, intuitive approach is called the *partial sum algorithm*. Students should be able to move through these algorithms to the adult algorithm <u>at their own pace</u>.

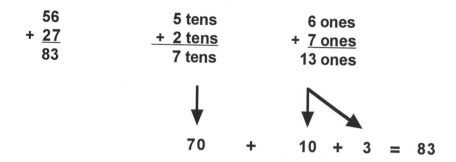

$$
\begin{array}{r}
56 \\
+\ \underline{27} \\
83
\end{array}
\qquad
\begin{array}{r}
5\text{ tens} \\
+\ \underline{2\text{ tens}} \\
7\text{ tens}
\end{array}
\qquad
\begin{array}{r}
6\text{ ones} \\
+\ \underline{7\text{ ones}} \\
13\text{ ones}
\end{array}
$$

$$70 \quad + \quad 10\ +\ 3\ =\ 83$$

The same principle is true of the adult multiplication algorithm. A child might multiply as illustrated below, called the *partial product algorithm*. Both of these algorithms work because of the distributive property.

$$
\begin{array}{r}
56 \\
\times\ \underline{27} \\
\end{array}
$$

$$
\begin{array}{lcr}
7 \times 6 & \longrightarrow & 42 \\
7 \times 50 & \longrightarrow & 350 \\
20 \times 6 & \longrightarrow & 120 \\
20 \times 50 & \longrightarrow & \underline{1000} \\
& & 1512
\end{array}
$$

C. NCTM position

K-4

Increased attention should be paid to the meaning of operations and a development of "operation sense" in students.

Students will do mental computation and estimate the reasonableness of answers.

Students will select appropriate computational methods and use calculators for complex calculations.

Teachers should help students develop thinking strategies for learning basic facts.

Isolated treatment of tedious paper-and-pencil computations should be decreased.

Long division should be deemphasized, as should the exclusive use of rounding to estimate.

5-8

Exploring the relationship between whole number and rational number operations should be emphasized, as it is important for understanding ratio, proportion, and percent.

9-12

Uses of operations are further explored in the contexts of algebra, functions, trigonometry, statistics, probability, geometry, and discrete mathematics.

D. Integrating problem solving, reasoning, communicating, and making connections

Students should be actively involved in problem solving and connecting mathematics to real-world situations, and in connecting the variations of all of the operations to the algorithms. Students should be encouraged to communicate their solutions and algorithms in mathematically meaningful ways through writing and oral discussions. These opportunities to share will solidify their understandings. When students choose appropriate operations to solve problems, they are reasoning.

E. Prerequisite skills and knowledge

Knowledge of numbers and a well-developed number sense are prerequisites for successfully understanding number operations. To add five to another number you must understand the concept of five or "fiveness," which can also be called numerosity.

Gelman and Gallistel (1978) wrote that to really understand counting, a child must understand five principles: one-to-one, stable order, cardinal, abstraction, and order-irrelevance.

1. *One-to-one principle.* Each item in a set must be assigned only one label (numeral).

2. *Stable order principle.* Numbers must be ordered appropriately and consistently for each counting set.

3. *Cardinal principle.* The final word uttered is the number label that is assigned to a set, and it identifies the numerosity of the set. For example, in counting three objects, "one, two , three," the last number named is three so there are <u>three</u> objects.

4. *Abstraction principle.* Any set of objects can be counted.

5. *Order irrelevance principle.* It doesn't matter what numeric label is put with which object when counting the set. For instance the purple frog doesn't have to be "six" because my favorite color is purple and I'm six years old. The green frog could have the label "six" attached to it instead.

Place value is also important. Students must understand "trading" for equal values and learn that positions of numbers have meaning. They must also understand the place holder nature of zero. They must construct a multiunit conception of number so they can <u>understand the operation</u>, not just memorize an algorithm.

F. Students' difficulties, confusion, and misconceptions

Many situations create the need to do a specific operation. Students sometimes have problems choosing an appropriate operation for a given situation, as in a word problem. Teachers can help students by understanding the operations in more depth, using student terminology, and contributing a variety of synonyms.

Most computation errors are consistent and happen due to flawed algorithms (*e.g.*, Ashlock, 1998). Many of these faulty algorithms demonstrate a poor understanding of place value.

The order in which the numbers of a problem are placed can create confusion (Weaver, 1971). For example, in the number sentence $2 + 3 = 5$, most students don't have as much trouble with $2 + ? = 5$ as they do with $? + 3 = 5$. Likewise, $5 = 2 + ?$ is more difficult for students than $2 + ? = 5$.

Many students do not understand the consequences of a poor choice of units during multiplication and division. Brown (1981) asked children to create story problems for 9×3. One 12-year-old came up with the following problem: "Lee has 9 and Jim had 3 chocolates. If you multiply them how much do they have?" (Brown, 1981). The student needed to realize that chocolates don't multiply in this instance. A correct problem would be: "If a child bought nine boxes of chocolates for $3 each. How much money did the child spend?" Correct use of units will often lead a student to a correct answer.

G. Factors contributing to students' difficulties, confusion, and misconceptions

When students learn that "less" or "fewer" in a word problem means subtract, many problems will be solved incorrectly. For example, if the word problem is: "Susie has three fewer apples than Jared. If Susie has five apples, how many does Jared have?" the solution would be obtained by addition, not subtraction.

Key words, therefore, are NOT a solution (Thompson & Hendrickson, 1986). By teaching key words, we raise barriers to the understanding of the situations for students. When students are taught key words such as "more" or "less" to indicate addition or subtraction, they sometimes focus on the words to tell them what operation to do instead of interpreting the situation, which could imply a different operation. For example, Thompson and Hendrickson (1986) demonstrated that the key words could be used for a variety of operations depending on the phrasing of the problem.

The language of place value is confusing in English, as discussed earlier. The lack of standardized words when referring to tens ("-ty" and "-teen") can confuse children, who want words to conform.

Poor textbooks are another part of the problem. Fuson (1990) analyzed several mathematics text series and found that they taught the concept of multiunit addition and subtraction later than necessary because of the emphasis on unitary conceptual structures. The multiunit approach is important for the development of place value concepts. For example, developing number concepts through twenty would help with the addition concept because the numbers past ten (such as 15) could be perceived as ten plus five and be developed. Place value could be developed as a natural part of our number system

Flawed algorithms occur because students are taught algorithms to find the answers before the underlying concepts are introduced. This procedural learning without conceptual understanding of the operation is harmful. It increases the likelihood of later errors that are a result of remembering garbled algorithms and having no way to check for accuracy because of a lack of conceptual understanding.

H. Appropriate teaching strategies

The discrepant event teaching strategy forces students to explain their approaches and confront any problems with their constructed algorithms. For example, when a student has a misconception involving a constructed algorithm such as the one involving fractions shown below, the activity could be to work through the addition problem using concrete manipulatives representing a whole unit. The activity could revolve around the dilemma of inconsistent answers.

$$\frac{1}{5} + \frac{2}{3} = \frac{3}{8}$$

Working collaboratively helps students revise and understand their created algorithms. It provides an opportunity for students with faster, shorter algorithms and a conceptual understanding to teach others and refine their reasoning.

All teaching about algorithms and problem solving should involve students in discussing and demonstrating their work with operations. Conceptual understanding is enhanced when children discuss, demonstrate, and explain their ideas with peers. Classroom discussion is also an opportunity to review a variety of situations about the operations and to make the concepts more concrete in children's minds.

The use of invented algorithms to strengthen place value concepts is important

(Kamii, Lewis, and Livingston, 1993). Rather than simply override a student's method with the adult version, it is critical for teachers to understand these invented algorithms. In this way, they can help students develop and have *ownership* of their methods for doing mathematics. Once teachers are aware of how their students are doing the calculations, questions should be posed to make the students think about and explore their thinking about place value.

The "Acting out the situations" activity has students create mental models, so they might realize that the same operation can have different situations.

I. Teaching notes

Manipulatives can be either proportional or non-proportional. Any manipulatives where the size implies the worth are *proportional*. For instance, power of ten blocks, beans and cups, and Popsicle sticks are all proportional. *Non-proportional* manipulatives could include chip trading materials, money, and an abacus. These materials have their worth agreed upon. For instance we might agree that three blues equal 1 yellow or five pennies equal one nickel.

It is important to begin working with proportional materials and then switch to non-proportional materials. As with all manipulatives, use a variety so the students can better relate multiple concrete experiences to the concept.

After students have experienced the manipulatives and discussed what they have discovered, the links from the concrete to the pictorial and symbolic need to be made obvious by the teacher.

This linking is particularly important in the "Acting out the situations" activity. Children should be encouraged to act out the situations of the problems. Creating a series of concrete activities where the students solve the problems and discuss them in terms of the necessary operations helps students experience the variety of meanings of operations.

For example, notice how different the actions and ideas are between the two examples below.

> *Problem 1:*
> Maria has 5 pens and Angela has 3 pens. How many more pens does Maria have than Angela?
> *An action:*
> Line the pens up and compare the number of pens for $5-3=2$.
>
> *Problem 2:*
> There are 15 children in the classroom. If nine are boys, how many are girls?
> *An action:*
> Use a manipulative to represent each child, separate out nine, and count the remaining for $15-9=6$.

Additional approaches are represented in the following examples.

Problem 3:
I have fourteen baseball cards and I need twenty-five to complete this album. How many more baseball cards do I need?
An action:
Make fourteen and count on until you are at 25 for 25-14=11.

Problem 4:
You gave me three jelly beans. I already had two jelly beans. How many jelly beans do I have now?
An action:
Have two beans, put three more with them, and count them up for 2+3=5.

Problem 5:
Kareem has six papers to color. Roger asks him to share and Kareem gives him two of his papers. How many papers does Kareem have now?
An action:
The action would be to start with six papers, remove two, and count those remaining for 6-2=4.

All of these situations could be made easier or more difficult and applied to a variety of classrooms. What is really important is to connect the manipulative to the action in the problem. Also, don't take away the manipulative in a comparison subtraction problem. Leave it there to demonstrate what the problem is asking. Role-playing activities also can be done for the multiplication and division models in the exercise.

Polya's (1985) steps for solving any problem could be emphasized at an early age. Questions that Polya proposed for understanding the problem include: "What is unknown?," "What is known?," "What is the condition?," and "Can you separate out the condition?." To devise a plan, find a connection between what you know and what you need to find out. This plan is carried out in the third step. The fourth step is to check the result with the question to see if it makes sense and answers the question.

Activities 2 through 8 are games to develop the concept of place value. Many texts and series advocate the use of manipulatives to teach place value. Some popular choices include power of ten blocks, beans and cups, Popsicle sticks, and chip trading materials. These manipulatives are important in developing the concrete notion of value by position. The place value games can be played with both proportional and non-proportional manipulatives but children should have experiences with proportional manipulatives in the earliest stages.

J. Materials needed

power of ten blocks
chip trading materials
Unifix cubes
beans and cups
craft (Popsicle) sticks and rubber bands
dice or spinners
calculators
blank cards or paper
pencils, strings, and other items specifically called for

K. Activities

See the following pages.

1. Acting out the situations

As you respond to each situation (a-i) below, *use the manipulatives* to act it out. *Then* make a drawing based on what you acted out with the manipulatives to show what is going on in the problem. Also write any explanation you think is necessary to explain your actions and reasoning.

An example of acting out the situations:

The situation: Maria has 5 pencils and Angela has 3 pencils.

The problem: How many more pencils does Maria have than Angela?

Acting out the situation and drawing conclusions:

Step one: The pictures show what we are given, from which we need to find out how many more pencils Maria has than Angela.

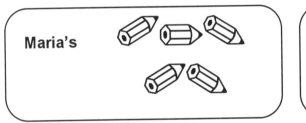 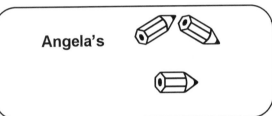

Step two: We decide to use a string to make a line connecting each of Angela's pencils to each of Maria's. The ones that aren't connected to another pencil are the extra ones that Maria has.

Step three: The action to carry out this plan is to line the pencils up and compare the rows of pencils. The action shows that Maria has two more pencils than Angela. One way to say it mathematically is 5 - 3 = 2.

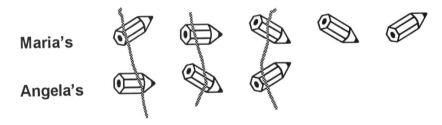

Step four: As we think about it, this answer <u>makes sense</u> because there are two more pencils and the number sounds reasonable for the original numbers given and the question being asked

---

Situations for students to solve:

a. There are 15 children in the classroom. Nine are boys, how many are girls?

b. I have fourteen baseball cards and I need twenty-five to complete this album. How many more baseball cards do I need?

c. You gave me three jelly beans. I already had two jelly beans. How many jelly beans do I have now?

d. Kareem has six papers to color. Roger asks him to share and Kareem gives him two of his papers. How many papers does Kareem have now?

e. Gary has more candy than he should eat. He decided to give some to his three friends and keep an equal amount for himself. If he gave each friend five pieces of candy, how much did he have to begin with?

f. It was Rachel's job to pick up the milk for her class. The milk cartons came in a flat where there were five cartons in a row and four cartons in each column. If the flat was full (so that all of these rows and columns were filled in with cartons) how many cartons were in each flat?

g. Julio was deciding what to get at the local ice cream store. He only had enough money for a single scoop of ice cream and there were three different kinds of cones: sugar, cake and waffle. If there were only five kinds of ice cream that he liked (strawberry, vanilla, chocolate, chocolate chip, and rocky road) how many different combinations of cone and ice cream did he have to choose from?

h. Grandma has $40 for her grandchildren's birthdays and five grandchildren. If she decided to give them all an equal amount for their birthdays, how much money should each child get? Do you think this is fair?

i. Warren has a board that is 15 feet long. He needs to make shelves that are 3 feet long. How many shelves will he get out of the board if each shelf uses the width of the original board?

2. Race to 100

The goal of this game is to reach 100 first.

To play the game:

> The first student rolls a die and receives as many units as are on the die face.
> When the student has 10 units, she/he trades to receive the tens piece (either a cup for 10 beans, a long for power of ten blocks, or a rubber band to put around the ten Popsicle sticks).
> The first student to receive the hundreds piece "wins."

3. Subtract to zero

The goal of this game is to begin with a hundreds piece and try to get to zero.

To play the game:

> The die is rolled and the student subtracts the number of units shown each time. When the student needs to regroup the 10 units, he trades and receive the tens piece or ones pieces that are appropriate.
> The first student to get to zero pieces "wins."

4. Banking game

The goal of this game is to reach a goal amount.

To play the game:

> One student acts as the "banker" and distributes the number of pieces each student needs.
> Each student rolls the die and tries to reach the goal piece by addition.

This game could be used for both base ten and other number bases if a teacher would like to try an alternative. The chip trading material should be used if other number bases are used.

The goal piece could be 100 or 1000 and two dice can be used to reach the larger numbers more quickly in base ten.

5. Calculator challenge

The goal of this game is to add on numbers to reach a total of exactly 50.

To play the game:
  One student begins by putting 1, 2, or 4 on the calculator. The next student adds to the number using a 1, 2, or a 4.
  The two students work together and add either 1, 2, or 4 to the sum that is shown on the calculator until one of the students gets exactly to fifty.
  The person who gets 50 is the "winner."

The students could then play with 1, 3, 5 and compare the strategies used for the two sets of numbers.

Variations of this game could include changing the target sum.

Also, the goal of the game could be changed to *avoid* reaching the number, adding 1, 2, 4 and 8, or 1, 3, 5, and 9. The person who reaches the number is the loser.

Another variation would be starting at a target and subtracting to zero

6. Card sharks

The goal of this game is to model place values with manipulatives, based on digits that come up on the cards.

You need two sets of cards with the digits from zero to nine written on them. Two to four players work well for this game.

To play the game:
  Each player chooses from each set of cards to create a two-digit number. This number should be modeled using the place-value materials.
  The cards are replaced in the set of cards. The next student should take a turn and model the number the student created with the cards.
  The next turn involves picking another set of cards and modeling that number.
  Then the numbers are added together doing the appropriate trading.
  The first student to reach 200 is the "winner."

It is important to let the students create the numbers with the digits because they will learn the strategy of using 93 rather than 39 when they form their number. This helps them with place value and an understanding of numerosity.

The set of cards should be referred to as digits, not numbers.

This same game could be played as a subtraction from 200.

7. Math BINGO!

The goal of this game is to build a BINGO! across a multiplication matrix.

Four dice can be used with two normal 1-6 dice plus two dice having the numbers 4-9. An empty multiplication fact chart can be created with graph or grid paper.

To play the game:
> The student rolls two dice of her/his choice and fills in the product that appears on the dice. For instance if a five and a six appear, then the student can fill in the five times six or the six times five slot.
> A variation would be to let students fill in both choices to reinforce the commutative property.
> BINGO! is achieved when the student has a row, column, or diagonal filled.

The strategy of "winning" this game should be explored with the students.

8. Secret numbers

The goal of this game is to use reasoning to identify a secret number.

Multiplication and division can be explored using the Secret Number. Calculators and reasoning should be a part of this game.

To play the game:
> The teacher or one student thinks of a number and gives clues such as
> > I'm thinking of a secret number that is less than 30.
> > This number is divisible by 5 with a remainder of 2.
> > This number is divisible by 11 with a remainder of 0.
> > What is my secret number?
> The students work to discover the number.

L. Assessment ideas

Many of the games used above can be a form of assessment. As students are busy playing the games or doing the tasks, the teacher can use their comments as anecdotal information for portfolios.

In putting together portfolios, students make choices about what to include to demonstrate their understanding and explain their choices.

Strategies used to "win" the games can be written in a journal, revealing information about what the student understands about operations and place value.

Another way to do a pencil and paper test is to ask the students to write their own problems. The teacher provides the operation sentence and the students write the word problem and solution strategy. This approach can tell teachers what operations and algorithms the students understand, as opposed to typical "right answer" tests.

Having students write their own word problems in class and solve each other's problems can be used to assess understanding. The student-generated questions also can be evaluated by other students and lead to discussions that reveal understanding and misunderstanding that will help the teacher adjust instructional experiences.

M. Resources and references

Ashlock, R. B. (1988). *Error Patterns in Computation, 7th Ed.* Upper Saddle River, NJ: Prentice-Hall.

Brown, M. (1981). Number operations. In K. Hart (Ed.), *Children's Understanding of Mathematics: 11-16.* London: John Murray.

Fuson, K. (1990). Issues in place-value and multidigit addition and subtraction learning and teaching. *Journal for Research in Mathematics Education, 21* (4), 273-280.

Gelman, R., & Gallistel, C. R. (1978). *The Child's Understanding of Number.* Cambridge, MA: Harvard University Press.

Kamii, C., Lewis, B. A., & Livingston, S. J. (1993). Primary arithmetic: Children inventing their own procedures. *Arithmetic Teacher, 41* (4), 200-203.

National Council of Teachers of Mathematics. (1989). *Curriculum and Evaluations Standards for School Mathematics.* Reston, VA: NCTM.

Polya, G. (1985). *How to Solve It.* Princeton, NJ: Princeton University Press.

Thompson, C. S., & Hendrickson, A. D. (1986). Verbal addition and subtraction problems: Some difficulties and some solutions. *Arithmetic Teacher, 33*(7), 21-25.

Weaver, J. F. (1971). Some factors associated with pupils' performance levels on simple open addition and subtraction concepts. *Arithmetic Teacher, 18,* 513-519.

A. Overview

Concepts included in this chapter are: importance of logic, situations which involve logic and reasoning, deductive and inductive reasoning, propositions, arguments, conditional statements, and Venn diagrams.

B. Background for the teacher

Why study logic?

Logic is the backbone of reasoning. As the NCTM Standards state, "Mathematics *is* reasoning. One cannot do mathematics without reasoning." Children need to make sense of mathematics, not just memorize rules and procedures.

Virtually all problem-solving situations require the use of logic. Scientists and engineers use logic in searching for patterns in nature, designing experiments and tests, and interpreting observations and data. Lawyers use logic in making arguments. Mechanics, plumbers, electricians, doctors, and other health-care professionals use decision-making protocols. Many computer applications use logic; *e.g.,* using the conditional "If-then" statement in programming or conducting a key word search.

What is the difference between deductive and inductive reasoning?

Deductive reasoning begins with a proposition, then uses a series of arguments from which the conclusion naturally follows. For example:

> Mathematics is fun.
> I like having fun.
> Therefore, I like doing mathematics.

A conclusion will always be true *if* all of the statements justifying it are true. Ultimately, the validity of any mathematical assertion is established using deductive reasoning.

Mathematicians value *elegance* in a deductive proof; that is, presenting a proof in its cleanest, most straightforward way. However, in doing proofs, elegance is one of those hard-to-define but easy-to-recognize aesthetic properties. Mathematicians strive for elegance in deductive proofs of mathematical assertions.

Inductive reasoning arrives at an answer through the exhaustion of all other possibilities. It begins with data (observations) rather than with a proposition. The data is examined to try to find a pattern. Then, the pattern is tested to see if it holds for predicted outcomes. For example: "The sum of the interior angles of every quadrilateral I've ever measured is 360°. So, I conclude that every quadrilateral has a sum of 360° for the interior angles."

Much of scientific research uses inductive reasoning. A conclusion using inductive reasoning is always subject to being proven false (falsification) if new, contrary evidence is found. The more evidence that seems to support an inductive proof, the more confidence we have in its correctness. Because of its openness to falsification, an inductive proof is a practical conjecture unless all of the possibilities are exhausted. Such certainty is probably impossible in a real-world environment. Inductive reasoning is being used more frequently in mathematics since computers can run exhaustive algorithms which may be able to dismiss all competing cases.

Propositions and logical arguments used in reasoning

Propositions are statements that make a distinct claim. For example: "All girls are wonderful." "No child may swim without an adult present." Propositions are the building blocks of arguments. A proposition can make an *assertion* or a *denial*. For instance, "Children should be seen and not heard." is an assertion, while "I did not wreck the car." is a denial. All propositions are declarative statements, having a subject and a predicate.

Four arguments can result from a proposition.

1.   *Negation.* A negation denies the truth of something. For instance, "The tile is not red." The tile could be any color but red. Another way to say this is, "It is not true that the tile is red."

2.   *Conjunction.* A statement that uses the connective <u>and</u> to combine two separate statements is a conjunction, such as "The tile is green and a square." Both original statements must be true for the conjunction sentence to be true. If the tile is not green, or has another shape, then the entire statement is false.

3.   *Inclusive or.* This argument is more complex than the previous two concepts. It is one of the ways that we use the connective <u>or</u>. "The tile is green or square." This could mean one of three things: 1) the tile is green, 2) the tile is square, or 3) the tile is both green and square. Another way to write an inclusive or statement is to use <u>and/or</u>, meaning that only one part of the sentence has to be true for the sentence to be true. If both parts are false, then the sentence is false.

4.   *Exclusive or*. This is the other way the connective <u>or</u> can be used. With this kind of statement, one or the other part must be true, but both cannot be true. For example, "The tile is a square or a hexagon" is an exclusive or statement, because one tile cannot be both shapes simultaneously.   One part of an exclusive or argument must be true for the statement to be true.

How do you tell between *inclusive or* and *exclusive or* propositions?

Since it is not always obvious which kind of "or" statement people are using, you might have to ask questions about the speaker's meaning. If someone were to tell you that you could go to the movies or go swimming, you would want to find out if you could do both. In mathematics, it is more obvious. "A number greater than ten or less than five" would imply an *exclusive or* because a number could not be both, as shown below.

By contrast, "A number greater than five or less than ten" could mean any number, an *inclusive or* situation, as shown below.

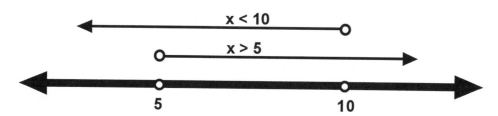

Don't get an *inclusive or* situation mixed up with the conjunction, "a number greater than five and less than ten" which would only be the numbers between five and ten. By convention in logic, you should always assume an "or" is inclusive unless it is stated otherwise.

What are conditional statements?

A conditional statement uses the <u>if-then</u> form. A conditional statement is often expressed generically as *If p, then q*. This means that if *p* is true it implies *q*. Use of the word <u>if</u> leaves the fundamental assumption of the statement open to testing and disproof.

Several forms of conditional statement can be made, negating and/or switching the hypothesis and conclusion, as shown in the following table.

STATEMENT	LOGICAL ARGUMENT	EXAMPLE
Original	If p, then q.	If I have a temperature of 101°, then I have a fever.
Converse	If q, then p.	If I have a fever, then I have a temperature of 101°.
Inverse	If not p, then not q.	If I do not have a temperature of 101°, then I do not have a fever.
Contrapositive	If not q, then not p.	If I do not have a fever, then I do not have a temperature of 101°.

The original and contrapositive statements for this example are both true. The inverse and converse statements are not true because my temperature could be 103° and I would have a fever. If the original statement is true, then the contrapositive will always be true, but the other statements are not always true.

Remember, the veracity of the original statement is always open to question, but if it is true or presumed true, then the contrapositive is also true.

Sets and Venn diagrams

In mathematics, a *set* is a collection of like objects that can be named. For instance, a table, chairs and sideboard might be included in a set called dining room furniture. We use sets every day when we attempt to group like objects. The objects in the set are members of the set. A *subset* is a part of the original set. Using our example, the chairs are a subset of the set of dining room furniture.

Venn diagrams are a graphic way to clarify and quickly explain situations. They use three different configurations of circles to demonstrate relationships. For instance, in the diagram below, the objects in A are all in B. Therefore, B contains every object in A. A is a subset of B.

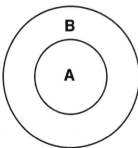

In the second diagram, only some of the objects in A are in B. The *intersection* of the two sets is comprised of the overlapping parts of the two circles where the objects are common to both. The shaded portion of the Venn diagram below represents those parts that are in both A <u>and</u> B.

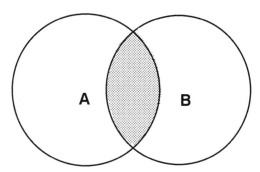

Disjoint sets are sets in which the members of the set have no commonalities. For instance, mammals and reptiles are distinct sets; there isn't a common member of both sets.

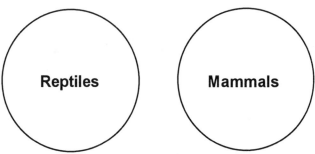

When two statements are logically equivalent, their Venn diagrams are the same.

C.   NCTM position

Mathematics as Reasoning is a major strand in the NCTM Standards that deal with the teaching and learning of logic.

K-4

Reasoning should be emphasized.

Students will draw logical conclusions about mathematics.

Students will use models, known facts, properties, and relationships to explain their thinking, and justify their answers and solution processes.

Students will use patterns and relationships to analyze mathematical situations.

Students will come to believe that mathematics makes sense.

5-8

Reasoning is expected to permeate the curriculum.

Students will recognize and apply deductive and inductive reasoning.

Students will understand and apply reasoning processes.

Students will make and evaluate mathematical conjectures and arguments, and decide the validity of their own arguments.

Students will appreciate the pervasive use of reasoning in mathematics.

9-12

Students will make and test conjectures and formulate counter examples.

Students will follow logical arguments and construct valid arguments.

Students will understand and construct simple proofs involving deductive and inductive reasoning.

Students are expected to do this sort of reasoning in many mathematical situations but not necessarily in the two-column proof format.

D. Integrating problem solving, reasoning, communicating, and making connections

Logic is a natural connection among disciplines. Connecting mathematics with art using the four-color map problem, for instance, requires the use of inductive reasoning. Logic is an important part of problem solving and is the language for reasoning an argument. Logical arguments are essential for communicating.

E. Prerequisite skills and knowledge

Most teaching and learning of logic does not have prerequisite skills and knowledge; however, the student must be mature enough to accept and evaluate arguments.

F. Students' difficulties, confusion, and misconceptions

Many teachers do not deliberately teach logic in any meaningful way. Much of what children come to understand about logical reasoning happens very indirectly.

There are more than twenty meanings for the word *set* as a verb and at least eighteen definitions as a noun. The mathematical meaning of the term (a group of elements that have a common characteristic or rule) must be clearly learned.

Many uses of the word *or* do not imply the meaning of the use. The use of an *exclusive or* to demonstrate a concept may be perceived as an *inclusive or* idea.

When diagraming the real number system, many student misconceptions about Venn diagrams can be found. For instance, some students have the idea that you can represent irrational numbers around the rational numbers and that the members of a set can be represented by points or dots. A Venn diagram is an area representation and does not include dots. It is continuous, not discrete.

G. Factors contributing to students' difficulties, confusion, and misconceptions

Many misconceptions arise because logic is not included in the formal mathematics curriculum in the early grades. Most textbook series ignore logic until high school geometry. The teaching of logic does not have to address formal logic, but the ideas presented should be an important part of early mathematics instruction.

H. Appropriate teaching strategies

The Venn diagram is a powerful tool for understanding logic and can be used with mental model building and other strategies. Using discrepant events helps students to test their ideas, forcing them to explain their reasoning and confront any problems with their constructed arguments. Venn diagrams help students to check if a discrepant event can occur.

Collaborative learning strategies help students revise and understand their created arguments, and can serve as a way for students with more advanced arguments to teach others and refine their reasoning. Sharing Venn diagrams allows students to depict and defend their logic.

All teaching about logic should involve students in discussion and demonstration of their work using Venn diagrams. For example, many discussions should occur during the "Sorting and classifying" activity so teachers can understand the children's labels of the Hula hoops and, therefore, their understanding of the concepts. Discussing, demonstrating, and explaining help children clarify their logic, as does the Venn diagram.

I. Teaching notes

Instruction in logic in our schools has often been haphazard. For some time geometry instruction was supposed to promote logic because of the heavy emphasis on proof; but doing a formal two-column proof does not mean that students understand logic or understand how to reason. Most geometry teachers have encountered the student who believes that all proofs look alike and many algebra teachers can attest to the lack of transfer from geometry to proof situations in algebra.

When teaching Venn diagrams, be sure to look for the appropriate labels and meanings. The questions you pose about Venn diagrams are important. Be sure to ask students why the labels work.

J. Materials

> attribute blocks
> Hula hoops or large circles of yarn/string (different colors help)
> checkerboards
> a variety of other games involving logic for winning strategies

K. Activities

See the following pages.

1. Sorting and classifying

Ask students to sort and classify attribute blocks, using Hula hoops to create Venn diagrams. Give the Hula hoop orientation and ask the groups of students to label and put the attribute blocks into the Venn diagrams (Hula hoops). The main issue at this point is the impossibility of using the same characteristic (*e.g.,* color) to describe both circles in an overlapping diagram. Students should be able to clearly define the characteristics of the pieces they are sorting, become comfortable with the materials, and recognize the characteristics for sorting (size, color, shape, and possibly thickness).

Next, ask the groups to do a one-difference train that loops back onto itself, with every piece having one different characteristic from the next and two that are the same. It helps to have the children say the characteristics (*e.g.,* "same color, same size, different shape") as each student takes a turn to establish the one-difference, two-of-the-same pattern. Since the goal of this lesson is to have students think about patterns and strategies, promote collaboration and discussion throughout the activity. They will see there is not just one correct answer. After the students have completed the trains, ask "What strategies did you use to solve this problem?"

Next, ask the students to create a two-difference train, sharing their strategies with others in their group, then sharing and discussing their patterns and strategies as a class. Finally, ask students to create a three-difference train.

After all of the trains have been completed, have students identify which sortings were more difficult and what makes them sorting easier or more difficult.

2. Let's diagram it!

Use the best fitting Venn diagram (see three samples below) to portray each of the given statements. Be sure to label your diagrams.

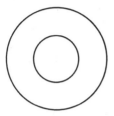 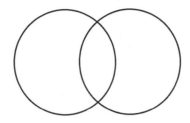 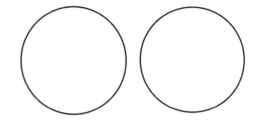

Example: All bats have wings.

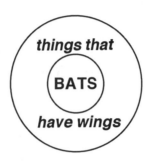

Some statements to diagram:

a.   All snakes are reptiles.

b.   Some desert animals are snakes.

c.   No polar bears are reptiles.

d.   No reptiles are mammals.

e.   All mammals are vertebrates.

f.   A turtle is a reptile.

g.   Some reptiles crawl on their bellies.

h.   There are reptiles that have short legs.

i.   Whales are mammals.

j.   There are no invertebrates that are vertebrates.

3. Analyzing games

a. Tic-Tac-Toe

Guide the students through a sequence such as this:

Play Tic-Tac-Toe with a partner and figure out a strategy for the placement of the "X" that will always win or cause the game to be a draw. What is the strategy you predict will always win?

Try several games with a new partner to test the strategy. Refine as needed. Explain why it works. Apply the strategy by explaining where you should place the initial "O" and why. Explain a strategy that you would use to win the Tic-Tac-Toe game.

b. Checkers

Predict a strategy for winning at the game of checkers. Play the game several times to consider some ideas. What strategy did you and your partner decide might work? Participate in a classroom discussion about strategies. Then play two games of checkers with your partner. Try to analyze the strategy you predicted to see if it ensured a win for this game.

How did your strategy work?

Why did it work (or why did it not)?

Do you have any new predictions for a strategy for winning checkers?

Discuss these new predictions and strategies with the class. Try them out with a partner. Explain your strategy for winning at checkers.

c. Other games

Explain a game that you enjoy playing. What strategy do you predict will always work to win at this game? Try the game out with a partner. Does your strategy work? Why or why not? What is your new strategy for winning at the game?

4. Logic problems

Design your own logic problem and challenge others to solve it. You might want to begin at an easy level with two variables and progress to a more difficult level with more pieces of the puzzle to figure out.

Example:
Four students are interested in having you figure out their whole names. Their first names are Anne, Betsy, Chuck, and David.

1. Harridon is a girl's last name.
2. Betsy and Anne don't think you can solve the problem but Irving thinks you can. Chuck isn't sure if you can solve the problem.
3. James is one of the boy's last names.
4. Anne isn't King.

5. Classical conditionals

Write the converse, inverse, and contrapositive statements for each statement given and then find the truth possibilities for these statements. Assume the original statement is true.

Example:

Original statement:
>   If I go to the movies, then I will go to see *Batman*.

Converse statement:
>   If I go to see *Batman*, then I will go to the movies.

Inverse statement:
>   If I do not go to the movies, then I do not see *Batman*.

Contrapositive statement:
>   If I do not see *Batman*, then I will not go to the movies.

Truth:
>   If the original statement is true, then the contrapositive is true, because I will only go to the movies to see *Batman* and if I don't see *Batman* then I can't have gone to the movies. The converse and inverse statements might not be true because I might rent *Batman* on videotape.

Some statements for students to work with:

a.   If two lines are perpendicular, then the angle they form is a right angle.

b.   If a polygon has three sides, then it is a triangle.

c.   If a four-sided polygon has opposite sides equal in length, then it is a parallelogram.

d.   If the sum of the interior angles of a polygon is 180°, then it is a triangle.

e.   All rectangles have four 90° angles. (Put in "if p, then q" form.)

f.   You make up a statement.

g.   Borrow a statement from your partner.

L. Assessment ideas

Have students evaluate statements such as in the Classic Conditional activity. These statements can be created and checked by fellow students.

Have students create Venn diagrams and write about them in their journals.

Devise an activity along the lines of "Red Rover" to do a Venn diagram assessment for early learners, in which students write their names or stand in the appropriate spots for labels provided by the teacher; *e.g.*, "Students wearing blue," "Students wearing jeans," etc.

M. Resources and references

Bennett, J. O., Briggs, W. L., and Morrow, C. A. (1996). *Quantitative Reasoning: Mathematics for Citizens in the 21st Century*. Reading, MA: Addison-Wesley.

Davis, P. J. and Hersch, R. (1982). *The Mathematical Experience*. Boston: Houghton Mifflin.

National Council of Teachers of Mathematics. (1989). *Curriculum and Evaluation Standards for School Mathematics*. Reston, VA: NCTM.

Paulos, J. A. (1995). *A Mathematician Reads the Newspaper*. New York: Harper Collins.

4
PATTERNS AND RELATIONSHIPS

A. Overview

Concepts included in this chapter are: the importance of patterns, difference between relation and function, recognizing and using patterns.

B. Background for the teacher

Why are patterns important?

A *pattern* involves the repetition of two or more items or sets. Beginning with very young children and continuing into high school and beyond, searching for and establishing patterns and relationships are important concepts and skills in mathematics. Some educators believe that mathematics *is* the study of patterns. By learning to decode patterns, one is able to extend to beyond the pieces of the given pattern.

The study of patterns may start with a look at color, size, or shapes. In each case, children are initially encouraged to focus on the relationships among the elements of the specific pattern. The same process is true when students deal with more sophisticated patterns. Secondary students learn to write mathematical relationships to represent a pattern.

What is the difference between function and relation?

A function is a special pattern. The difference between a relation and a function can be explained in three statements:

> A *relation* is a set of ordered pairs.
> A *function* is a relation for which there is exactly one value of the dependent variable for each of the values of the independent variable.
> All functions are relations but not all relations are functions. (See Venn diagram, following.)

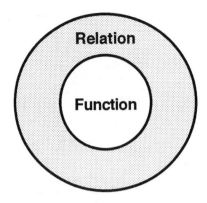

Examples of patterns

Below are some examples of activities which involve patterns:

What is the next number?
What is the next shape?
Guess the procedure being used.
Find the pattern in a table of data.
Use a pattern to find the sum of the first 100 odd numbers.
Find the pattern involved in a graph.
Use the pattern found in a table of data or a graph to make predictions.

C. NCTM position

K-4

Students will recognize, describe, extend, and create patterns.

Students will represent and describe mathematical relationships.

Students will explore the use of variables and open sentences.

5-8

In addition to the processes recommended for K-4, students will:

Describe and represent relationships with tables, graphs, and rules.

Analyze functional relationships to explain how a change in one quantity results in a change in another.

Use patterns and functions to represent and solve problems.

In addition to the above capabilities:

Students will model real world phenomena with a variety of functions.

Students will represent and analyze relationships using tables, graphs, equations, and rules.

Students will translate among tabular, symbolic, and graphical representations of functions.

Students will analyze the effects of parameter changes on the graphs of functions.

D. Integrating problem solving, reasoning, communicating, and making connections

As students work on searching for and establishing patterns and relationships, they should, whenever possible, use the steps involved in problem solving, be involved in reasoning, use a variety of ways to communicate mathematical ideas, and connect them to real-life situations.

E. Prerequisite skills and knowledge

Not all the activities included here are appropriate for students at all levels. Some of the introductory activities may be used with very young children. Some activities are designed for older students, and still others are more appropriate for students taking beginning algebra. The teacher should make sure that the students have the prerequisite skills and knowledge for the activity selected.

F. Students' difficulties, confusion, and misconceptions

Many students lack experience in searching for and establishing patterns. As a result, they may not have the necessary prerequisite skills for searching for and establishing patterns.

G. Factors contributing to students' difficulties, confusion, and misconceptions

Some of the most common factors that contribute to problems students have with the concepts of patterns and relationships are:

- Textbooks lack adequate attention to patterns.
- Rules and formulas are often presented without much discussion
- Instruction involves presentation of other peoples' generalizations without allowing students adequate opportunities to make sense of those generalizations
- Some problems and exercises are presented at a level of abstraction for which the students may not be developmentally ready.

H. Appropriate teaching strategies

Recommended strategies for these topics include teaching for conceptual change, discussion and demonstration, and discrepant events.

These strategies provide students the opportunity to make predictions, be active in the learning process, develop the skill of searching for and establishing patterns, and construct their own knowledge and skills in regard to patterns and relationships.

Discrepant events can motivate students to seek ways to solve the mystery and to find the pattern. Discussion and demonstration are important for giving students the opportunity to share their thinking as they struggle to search for patterns. Demonstrations by the teacher should not, however, take the place of actual hands-on and collaborative experiences for the students.

I. Teaching notes

The activities in this chapter call for a range of abilities and all may be used with secondary mathematics students. A few activities also are appropriate in elementary mathematics classes. Activities 1, 2, 3, 4, and 14, and--to a limited extent--10 and 12 may be used with both elementary and secondary students. Other activities are more appropriate for secondary students.

With younger students, you may begin with "What comes next" activities. Decide on a criterion; *e.g.*, glasses or no glasses, short sleeves or long sleeves. Ask a student with glasses to come to the front of the class, then ask another without glasses, then with glasses. Challenge students to find your pattern. After doing this several times, invite individual students to come up with their own patterns.

J.   Materials

25 small, triangular pattern block overhead projection pieces
sets of pattern block pieces
cylinders of various sizes
rectangular solids (prisms) of various sizes
rectangular solids of same heights but with square bases of different sizes
about 40 blocks of the same size
several Geoboards and rubber bands to use with them
colored papers
pendulum bobs
string
rulers
meter sticks
graph or grid paper

K.   Activities

See the following pages.

1. What comes next?

For each sequence below, predict what you think will be the missing numbers and explain your reasoning. Share your predictions with others in your group and then with others in the class.

a.	1	3	5	7	__	__	__
b.	1	4	7	10	__	__	__
c.	3	6	9	12	__	__	__
d.	1	2	3	5	__	__	__
e.	1	4	9	16	__	__	__

2. Predict the answer

Ask the students to follow along as you give them verbal instructions:

> Write a number.
> Add 9.
> Double the number.
> Subtract 4.
> Divide by 2.
> Subtract your original number.

Have them complete the calculations, then say: "I predict the answer is 7."

Then ask the students, "How did I do this?"

3. Putting shapes together

Make your own predictions for situations a and b below. Share your predictions and reasoning with others in your group and then with others in the class.

Situation a.

Here is the first square:

It takes 4 squares to make the 2nd full square:

It takes 9 squares to make the 3rd full square:

How many squares does it take to make the 4th full square?

How many squares does it take to make the 50th full square?

Situation b.

Here is the first triangle:

How many triangles does it take to make the 2nd full triangle?

How many triangles does it take to make the 3rd full triangle?

How many triangles does it take to make the 4th full triangle?

How many triangles does it take to make the 50th full triangle?

4. What is the sum of the 1st 100 odd numbers?

After the students work the first three activity sets, moderate a discussion about the patterns they have recognized, and how recognizing a pattern establishes a basis for prediction, such as the question posed here.

Help them to think of real-world connections and applications.

5. Ratio of circumference to diameter

Some people claim that regardless of what size cylinder we have, the ratio of the distance around the cylinder to the distance through the middle is always the same.

Do you agree? How can we prove or disprove this claim?

Share your views with others in your group. Select someone from your group to present the views of the group to the class.

How do your views compare with others in the class?

Are there some insights which you may borrow from others?

Get the necessary materials and test your ideas.

Present your data in the form of a table.

Do you notice a pattern developing?

Present your data in a graph.

Write a statement representing the pattern in your table and/or your graph.

How can you represent the relationship found in your table, graph or statement in the form of an equation?

6. Relationship between volume and height of a cylinder

Predict the relationship which exists if you plotted the volume of a cylinder and its height.

Why do you think so?

Share your predictions with others in your group. Select someone to present the predictions and explanations of the group to the class.

How can you test your predictions?

Get the necessary materials and test your predictions.

Based on your observations, do you want to make any changes in your explanation?

7. Relationship between volume and the area of the base

If we plot the volume of rectangular boxes or prisms versus the area of the base, what relationship do you think will result?

Share your predictions and reasons with others in your group.

Get the necessary materials and test your predictions. You will need at least 5 data readings.

Did your observations agree with your predictions?

What would happen if we used a cylinder instead of the box or prism?

Test your predictions and present the results to others in your group and class.

Expressed mathematically, what is the ratio of volume to the area of the base? What are the units?

8. Predict the values for the missing numbers

There is a relationship between A and B. You are given corresponding B values for 0, 1, 2, and 5.

What is the corresponding value of B for A = 50? for A = n?

A	B
0	1
1	2
2	5
--	--
--	--
5	26
--	--
--	--
--	--
50	?
--	--
--	--
--	--
n	?

9. Can you figure out the pattern for the polygons?

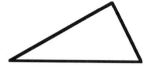 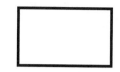

The sum of the internal angle measures of a triangle is 180 degrees.
The sum of the internal angle measures of a rectangle is 360 degrees.
The sum of the internal angle measures of a hexagon is 720 degrees.

What is the relationship between the number of sides of a polygon and the sum of the internal angle measures?

Using the pattern you have found, answer the following questions:

What is the sum of the internal angle measures of a 10-sided polygon?

What is the sum of the internal angle measures of an *n*-sided polygon?

10. Stacking blocks

The table below shows the relationships between the number of blocks and the height of the stack. You are given the first 5 readings. Using the pattern, give the number of blocks for a stack that is 10 high? for *n* high?

Height of stack	Number of blocks
1	1
2	3
3	6
4	10
5	15
--	--
--	--

11. Number of sides and diagonals of polygons

The table below gives the corresponding number of diagonals for polygons of various numbers of sides. Using the trend, find the missing number of diagonals for the polygons of given sides.

Number of sides	Number of diagonals
3	0
4	2
5	5
6	9
7	--
10	--
--	--
--	--
n	--

12. Painting cubes

The table below shows the dimensions of large cubes and the corresponding number of small cubes that make up the cubes, the number of cubes with one face painted, with two faces painted, etc. Two sets of cubes are shown as examples. Using the pattern, fill in the missing information on the table.

Large cube dimensions	# of small cubes in large cube	With one face painted: # of small cubes having paint	With two faces painted: # of small cubes having paint	With three faces painted: # of small cubes having paint
2 x 2 x 2				
3 x 3 x 3				
4 x 4 x 4				
5 x 5 x 5				
---				
10 x 10 x 10				

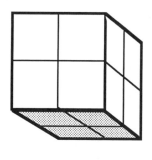 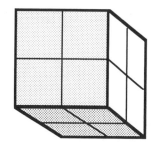

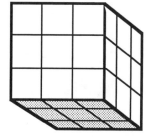 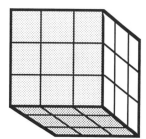

13. Is there a relationship between the period and the length of a pendulum?

The time it takes for a pendulum to make a complete swing, out and back, is called its *period*. The *effective length* of the pendulum is the length from the point of suspension to the center of mass.

What do you think is the relationship between the period and the effective length of the pendulum?

Why do you think so?

Share your ideas with others.

Get the necessary materials and as a group, make predictions and test your ideas by collecting appropriate data.

Based on the data you collected, does there seem to be a relationship? What statement can be made about the existence of a relationship between the period and the effective length?

How can we use these results? Can you think of real-life examples?

14. Your own example

Using numbers, shapes, or physical phenomena, think of a situation involving patterns.

Present it to others and challenge them to find the pattern.

L. Assessment ideas

Using written examples, ask students questions such as: What comes next? What is your reasoning? What is the pattern? Using the patterns, ask students to make predictions.

Using concrete materials such as pattern blocks, ask students: What is the next shape?

M. Resources and references

Center for Occupational Research and Development. (1991). *Patterns and Function.* Waco, TX: Author.

Erickson, D. B. (1991). Students' ability to recognize patterns. *School Science and Mathematics, 91*(6) 255-258.

National Council of Teachers of Mathematics. (1989). *Curriculum and Evaluation Standards for School Mathematics.* Reston VA: NCTM.

Petreshene, S. (1985). *Mind Jugglers.* New York: The Center for Applied Research in Education.

Stepans, J. I. (1995, February). The power of mathematics is in communicating, searching for, and establishing patterns. *Illinois Mathematics Teacher.*

Thompson, F. (1992). Geometric patterns for exponents. *Mathematics Teacher, 85,* 746-749.

5
NUMBER THEORY

A. Overview

Concepts included in this chapter are: number theory, even and odd numbers, factors, multiples, prime numbers, composite numbers, the number one, prime factorizations, least common multiples, and greatest common factor.

B. Background for the teacher

Some definitions

Number theory is the study of integers and their relationships. For example, a look at the relationships between 3 and 12 might reveal that 3 is a factor of 12, since 3 divides evenly into 12, 12 is a multiple of 3, 3 is an odd number while 12 is an even number.

An *even number* is any number that can be divided by 2 to obtain a whole number. The *odd numbers* are whole numbers that cannot be divided by 2 to obtain a whole number; so {2, 4, 6 ...} are even numbers and {1, 3, 5 ...} are odd numbers.

A *factor* is a number that divides evenly into another number. Three is a factor of 12 because 3 x 4 = 12. All of the factors of 12 would be {1, 2, 3, 4, 6, 12}. If there is a natural number a which can be multiplied by a natural number c to obtain a number b so that $a \times c = b$, then a is a factor of b. We can say that a divides b, and we mean that a "fits into" b by a natural number divisor.

A *multiple* is a number obtained by multiplying a number by other numbers. Twelve is a multiple of 3, because 3 x 4 = 12. The multiples of 3 would include {3, 6, 9, 12...}. If a is a factor of b, then b is a multiple of a. So if there are two natural numbers a and c, so that $a \times c = b$, b would be a multiple of a and c.

A *prime number* has only 1 and itself as factors, so a prime number always has <u>only</u> two factors. The number 3 is a prime number because 3 can be factored as {1, 3}. Other prime numbers are 5, 7, 11, 13, and 17.

Composite numbers have more than two factors. For example, 8 is a composite number because it has the factors of {1, 2, 4, 8}. So we could say that 8 has four factors. The numbers that have an odd number of factors also have a special name. They are called *square numbers* because a square with sides of 5 units has an area of 25 square units. 25 is a square number because it has 3 factors {1, 5, 25}.

Is the number 1 prime or composite?

The number 1 is not a prime number because it only has one factor (1) and prime numbers have two factors. The number 1 is not a composite number either, because composite numbers have more than two factors. So, the number 1 is neither prime nor composite.

Prime factorizations

Composite numbers can always be written as a product of primes. The product is called a *prime factorization*.

For example, the prime factorization of 12 is $2 \times 2 \times 3 = 2^2$.

You can use either a factor tree or a linear division method to find the prime factorization of a number.

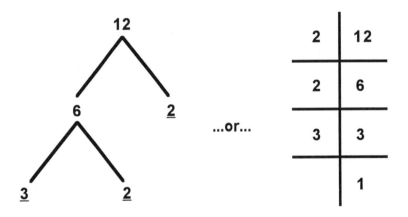

Least common multiple

The *least common multiple* (LCM) is the smallest number that two or more numbers can evenly divide so that when the division is performed there is no remainder. So 12 is the least common multiple for 3 and 4 because it is the smallest number that both 3 and 4 can divide evenly. To find the least common multiple, the multiples for both numbers can be figured until a common one is found:

3: 3, 6, 9, <u>12</u>
4: 4, 8, <u>12</u>

Another method is to identify the prime factorizations,

$$3: 3$$
$$4: 2^2$$

and build the LCM by using each prime factor the maximum number of times it occurs in either of the numbers. Thus, $2^2 \cdot 3 = 12$.

So, to find the LCM for 30 and 32, we could write the multiples:

$$30: 30, 60, 90, 120, \ldots$$
$$32: 32, 64, 96, 128, \ldots$$

which might take awhile, or we could find the prime factorizations:

$$30 = 2 \times 3 \times 5$$
$$32 = 2^5$$

These prime factorizations show that 2 is a factor of both 30 and 32, once as a factor of 30 and five times as a factor of 32. So 2 will have to occur five times as a factor in the LCM. Similarly, 3 and 5 are also necessary in the prime factors of the LCM. So the LCM will have to be $2^5 \times 3 \times 5 = 1980$. As you can see, prime factorization is a quicker method for large numbers. A Venn diagram clearly demonstrates this concept.

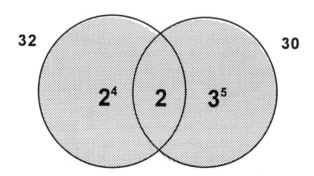

The LCM is useful for finding common denominators in fractions, such as the LCD (the least common denominator). For example, to add 1/2 + 1/3, you need to find a common denominator. The prime factorization of 2 is 1 x 2 and the prime factorization of 3 is 1 x 3. So, the LCD would be 2 x 3 = 6. One-half would be equal to 3/6 and one-third would be equal to 2/6. So 3/6 + 2/6 = 5/6. A common denominator is found so that we are referring to the same unit for all of the fractions. When we use decimals, our common denominator is tenths (or hundredths, thousandths, etc.).

Greatest common factor

The *greatest common factor* (GCF) is the largest number that will evenly divide into two other numbers. For instance, the GCF of 30 and 32 is 2 because 2 is the largest whole number that will "fit into" both 30 and 32. A Venn diagram can also demonstrate this concept, as shown below.

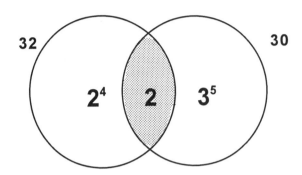

To find the GCF, use the prime factorization to find all of the factors common to both numbers. For example, to find the GCF for 18 and 24, we find the prime factorization for both numbers.

$$18 = 2 \times 3^2$$
$$24 = 2^3 \times 3$$

The factors that are common to both numbers are 2 and 3 so we multiply 2 times 3. Six is the GCF.

Why study number theory?

It is important to study number theory because it can be used in so many other parts of mathematics. If students understand divisibility rules, many computations can be made easier. Fractions, coding, and rational equations all use number theory. Mathematicians use number theory frequently, and there are many applications that can be found in security industries. For example, many codes during World War II were built using prime numbers. Computers are often made more secure using number theory concepts such as divisibility rules and prime numbers.

C. NCTM position

K-4

Students will understand our numeration system by relating counting, grouping, and place-value concepts, and developing number sense.

Students should develop these concepts using real-world experiences and physical materials.

5-8

Students will develop a concept of number systems.

Students will apply number theory concepts such as primes, factors, and multiples in real-world and mathematical problem situations.

9-12

Students will be able to compare and contrast the real number system and its various subsystems with regard to their structural characteristics.

All students will appreciate that seemingly different mathematical systems may be essentially the same.

College-intending students will develop the complex number system and demonstrate facility with its operations. Other mathematical structures such as groups and fields can be developed using axiomatic systems.

D. Integrating problem solving, reasoning, communicating, and making connections

Not only is number theory useful in solving many types of problems, but mathematicians study many number theory concepts with problem-solving methods. Problem-solving strategies help create new ideas about our number system. Number theory involves using reasoning to understand the relationships of numbers.

Number theory is critical when communicating mathematically in proofs. The reasoning involved in many mathematical proofs is supported by the underlying theory.

Problems that use number theory may have connections to computer programming and encoding systems.

E. Prerequisite skills and knowledge

Students must have an understanding of basic mathematics facts in addition and multiplication. Number theory is a way to further develop students' ideas about division so it would be useful to know basic division facts.

F. Students' difficulties, confusion, and misconceptions

Students often get confused about least common multiples (LCM) and greatest common factors (GCF). Many students think that if they find a least common multiple it ought to be smaller than the greatest common factor. Many of the confusions are a lack of understanding about which one they need to use for what kind of problem.

Prime factorizations and their importance for thinking about numbers may be ignored.

G. Factors contributing to students' difficulties, confusion, and misconceptions

The terminology may be counter-intuitive and contribute to confusion, since something that is "least" ought to be smaller than something which is "greatest."

Students are taught LCMs and GCFs without context or meaning, learning procedures to "Find the ......" instead of understanding the ideas. It begins to look like alphabet soup to a student who simply memorizes a procedure.

The importance of prime factorizations can be ignored by some students because they fail to realize the importance of number theory. The reasoning involved in number theory isn't sufficiently emphasized for some students to realize the importance.

Textbooks often present GCFs and LCMs within one page of each other or on the same page because they both involve solving use prime factorizations.

H. Appropriate teaching strategies

Discussion and demonstration is an excellent way to learn about students' understanding of number theory. In the "Factor mazes" activity, for example, students share their strategies for working the mazes. This sharing will help them construct better mazes on their own.

Mental model building is used in "Visualizing common numbers" and "Standing up for your numbers." The models created should help students understand what happens with number theory concepts.

I. Teaching notes

"Visualizing common numbers" is a concrete activity that will help students think about the properties that make up common numbers.

"Multiplication magic squares" asks students to consider number theory ideas such as "systems." The goal is to have students ask themselves questions about the concept of closure. A system has *closure* for an operation if the operation can be performed on two members of the system while still obtaining an answer within the system. For example, if I add two whole numbers, I still obtain a whole number which demonstrates closure for addition in the whole number system. The operation in the activity is multiplication. The system in the activity is number squares that work for multiplication.

"Multiplication magic squares" and "Factor mazes" reinforce prime factorization concepts, asking students to factor numbers and solve problems.

"Standing Up for Your Number" works best with at least 24 students. If there are fewer students, consider giving some students 2 numbers.

Be sure to practice the card turns for the factors in the activity "Standing Up for Your Number." Students need to realize that the cards should not turn for any number greater than the number they are holding. If they turn the card for a number past their number, it is often because they are confusing the term *factor* with *multiple*. This activity is appropriate for at least junior high students who can abstract the concept. Students should have concrete experiences with the concepts being explored prior to this activity.

J. Materials needed

> Cuisenaire Rods
> number cards
> graph or grid paper

K. Activities

See the following pages.

1. Visualizing common numbers

Place 3 purple rods so they are end-to-end lengthwise. Then use other rods to visualize the same situations.

<u>An example of visualizing common numbers</u>

The situation: Three purple rods are placed end-to-end.

The problem: How many light green rods can be placed end-to-end beside the purple rods to begin and end at the same place as the purple rods?

Visualizing the situation and drawing conclusions:

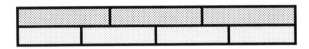

By using the manipulatives, it becomes apparent that it will take four green rods.

<u>Situations for students to solve</u>

a.   How many red rods would you have to put down to begin and end even with the purple rods?

How many white rods would you have to put down to begin and end even with the purple rods?

If the white rods were equal to 1 unit...

   What would the red rods be worth?

   What would the light green rods be worth?

   What would the purple rods be worth?

If it takes 6 red rods to equal in length four light green rods, how many white rods would be equal?

Show how these numbers are working.
      6 x _____ = 3 x _____ = 1 x _____.

How many white rods are needed to make these combinations finish at the same place?

The <u>least</u> common multiple is represented when all three meet in the same place for the first time. It is the smallest number that 2, 3, and 4 all divide. So what is the least common multiple represented by the red rods, purple rods and light green rods?

b.   Try some different rod combinations in your group.

What would be the least common multiple (the first time both end at the same place)

        for the yellow rod and the red rod?

        for the blue rod and the light green rod?

        for the purple rod and the brown rod?

        for the dark green and the purple rod?

        for the purple rod, the brown rod, and the orange rod?

c.   Make up a new combination of different color rods and find the least common multiple.

2. Standing up for your numbers

This activity is designed for the teacher to facilitate and to discuss the concepts with the class (discussion and demonstration). It is not intended as a worksheet.

Give each person in the class a card with a number from 1 to the size of the class. The cards should be large enough to read across the room. Arrange the students in a horseshoe formation, in numerical order. Have the students hold the cards face out so everyone can read the number.

Have a student demonstrate how to raise a card above his/her head. This is called "raising the card."

Then play:
   If you have an odd number, raise your card.
   If you have an even number, raise your card.

Say:
   "All numbers are either odd or even."
   "Think of all of the factors of your number and tell your neighbor what they are."

Have a student demonstrate a card turn. This happens when a student turns the card from either face forward to face backward or vice versa. The card does not turn completely around at a given time.

When I call out a factor of your number, turn your card. For example, if I have the card numbered 2, I would turn my card face backward when the number one was called out, and then I would turn my card face forward when the number two was called out. I wouldn't turn my card for any other numbers because only 1 and 2 are factors of 2.

Start counting out the numbers [1, 2, 3, 4, ...] slowly. Have students watch each other as the cards turn. Stop counting when you get about three past the number of students in the room.

Have the students look for patterns.

Some of the patterns that students will recognize include:

   • Square numbers are facing backwards because they have an odd number of factors.
   • The number of numbers between the square numbers forms a pattern.

Have the students turn all of their cards around again. This time have them keep track of the number of times they turn their card. For instance, the factors of 6 are {1, 2, 3, 6} so the student with the 6 card would turn her/his card four times. Count out the numbers again.

Tell all of the students who turned their card more than twice that they can sit down. (This should leave you with all of the prime numbers and one.) Ask the students who turned their card only twice to raise their card. Ask the students what these numbers are called. Ask the students what the numbers that sat down are called (composite numbers)

Let the prime numbers sit down. What's left? (One). Is one either prime or composite? (No, because it only has one factor. Prime numbers have two factors and composite numbers have more than two.) Collect the card with number one and replace it with a different number.

Now let's work with multiples. If I call out a multiple of your number, raise your card. Start counting slowly. When you reach four, stop and notice what numbers are raised. Is four a multiple of two? So what would be a common multiple for 2 and 4? So what is the smallest number that both 2 and 4 fit into?
Continue on with the counting and these types of questions.

3. Factor mazes[1]

Multiply the maze path numbers used to obtain the exit numbers. You may not move diagonally.

Example:
For the following maze, if I were to begin at the 2, what squares would I have to pass through to multiply and equal 27,720?

It might help to know the prime factorization for 27,720.

The prime factorization will tell me what squares I need to pass through.
$$27,720 = 2^3 \times 3^2 \times 5 \times 7 \times 11$$

So if I start at the 2, and I know that I need to pass through a 2 to get to the exit, then I only need one more 2. I also need to pass through at least two 3's. Since there is only one 3, I will have to pass through the 6 which will help me with a 2 and a 3. I know that I can't pass through three 2's and the 6 so I have eliminated one of the beginning 2's. I also know that I may not pass though both 11's. I need to find a path that will meet all of those conditions.

I'll trace one path. You find another.

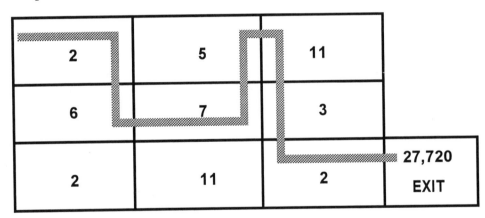

[1] Adapted from: Fitzgerald, W., Winter, M.J., Lappan, G., and Phillips, E. (1986). *Factors and Multiples*. Menlo Park, CA: Addison-Wesley.

Try another:

2	5	2	2	
11	2	11	5	**15,400 EXIT**
5	7	3	7	

Make a maze that has an exit number of 13,860.
What is your strategy for creating this maze?

				13,860 EXIT

Make a maze that goes through 9 squares before the exit.

				EXIT

Have your partner solve the maze while you solve the maze creation. What strategies did you use to make your maze?

4. Multiplication magic squares[2]

Multiplication magic squares have row, column, and diagonal products that are all equal. Is the array below a multiplication magic square? Explain your answer and the processes you used to get it.

1	4	2
4	2	1
2	1	4

Complete the multiplication magic square on the left below. What is the product of each row, column, and diagonal?

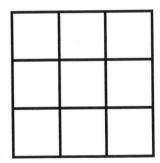

9	1	
	3	
		1

In the blank format on the right above, construct a magic square with a product of 64 for each row, column, and diagonal.

Explain the pattern you use for creating a 3 x 3 multiplication magic square.

[2] Adapted from: Sawada, D. (1975). Magic squares: Extensions into mathematics. In S. E. Smith & C. A. Backman (Eds.) *Games and Puzzles for Elementary School Mathematics*. Reston, VA: NCTM.

Take the magic square you created above and try multiplying every cell in the square by 2, using the format shown below. Is the result still a magic square?

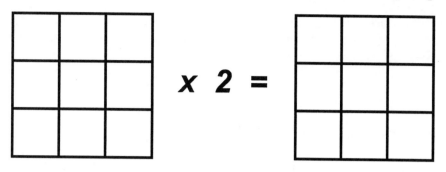

What prediction can you make about multiplying by other, different numbers? Try it, using the format below. Explain your observations.

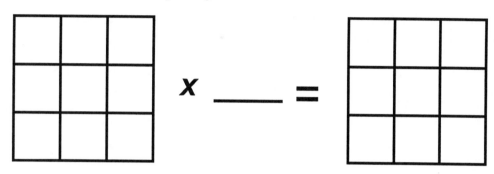

Try multiplying the two magic squares below. Multiply the first cell by the first cell, the second cell by the second cell, and so on, as started below. Complete the multiplication. Is the resulting square a magic square?

1	4	2
4	2	1
2	1	4

X

2	4	1
1	2	4
4	1	2

=

2	16	2
4		

If it is, the property you are demonstrating is called *closure*.

Do you think that if you multiply any two multiplication magic squares you always get another multiplication magic square? Test your prediction by multiplying two other magic squares.

Depending on your results, demonstrate why your results will always or never work when multiplying two magic squares.

L. Assessment ideas

Ask students to record in their journals such tasks as: Write the 7 digits of your phone number, choose four and arrange them to create a four-digit number. Find the prime factorization for your number. Tell about all of the special attributes of your number.

M. Resources and references

Fitzgerald, W. , Winter, M. J., Lappan, G. and Phillips, E. (1986). *Factors and Multiples.* Menlo Park, CA: Addison-Wesley.

National Council of Teachers of Mathematics. (1989). *Curriculum and Evaluation Standards for School Mathematics.* Reston VA: Author.

Sawada, D. (1975). Magic squares: Extensions into mathematics. In Smith, S. C., & Backman, C. A. (Eds.). *Games and Puzzles for Elementary School Mathematics.* Reston, VA: NCTM.

6
TWO-DIMENSIONAL GEOMETRY

A. Overview

Concepts included in this chapter are: the notion of covering, units, estimating, scale, figures, making shapes from other shapes, transferring, properties of 2-D figures, examples of 2-D geometry, history of 2-D geometry, who uses 2-D geometry, Pythagorean theorem, quadrilaterals, terms (congruence, similar, etc.), and symmetry.

B. Background information for the teacher

Examples of two-dimensional figures

A *rectangle* has four sides with opposite pairs being equal. A *square* has four sides of equal length and all four angles are right angles.

rectangle   square

A *triangle* has three sides. A triangle can be *isosceles, equilateral,* or *right*. An isosceles triangle has two equal sides. In an equilateral triangle all three sides have equal measures. In an *acute* triangle all interior angles are less than 90 degrees, whereas in an *obtuse* triangle one of the interior angles is larger than 90 degrees. A triangle with one right angle is called a right triangle.

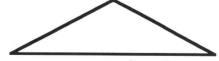

isosceles (and obtuse) triangle

equilateral (and acute) triangle  right triangle

A *parallelogram* has four sides, with opposite sides being equal and parallel. The term can be applied to a figure that has two acute angles and two obtuse angles, although a rectangle and square also fit the definition. A *rhombus* is a parallelogram having equal sides (like a square) but with two acute angles and two obtuse angles.

A *trapezoid* has four sides,with only two sides being parallel.

parallelogram   trapezoid

A *circle* is a set of points that are all equidistant from the center.

An *ellipse* is somewhat like a circle, but appears stretched along one of its diameters.

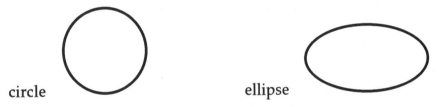

circle   ellipse

Units for measuring attributes of 2-dimensional figures

Three different measures are associated with 2-D figures: perimeter, angle, and area. *Perimeter* is the length around a shape, and is expressed in the same units as those for length--cm, m, in, foot, etc.. *Angles* are measured in degrees. The angle made by two perpendicular lines is 90 degrees. A circle is formed by a rotation through 360 degrees. The *area* of a 2-D figure is a measure of <u>coverage</u>. To measure area, we use square units--sq cm (cm^2), sq in (in^2), sq m (m^2), sq. ft (ft^2), etc.

How can we make shapes out of other shapes?

If we cut a rectangle, square, parallelogram, and trapezoid along their diagonals, triangles result.

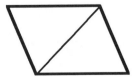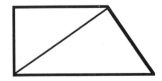

Making shapes from other shapes also help us make sense of the areas of different shapes. A *tangram* is an interesting puzzle that can be used to help students work on making shapes out of other shapes.

Who uses 2-D geometry?

Two-dimensional geometry has many practical applications. A builder who builds a house, a carpenter, a painter, a carpet layer, a machinist, a dress maker, a farmer, among others make use of and rely greatly on the geometry of 2-D shapes. Deciding how much fencing to buy and how much fertilizer to apply to a lawn requires an understanding of 2-D geometry.

Examples of 2-D shapes in nature

The surfaces of leaves, flowers, patterns on the wings of a butterfly, surfaces of mineral crystals, and shapes of snowflakes are just a few examples of how nature expresses the beauty of 2-D dimensional geometry.

Measurements used with 2-D figures

To fully describe a 2-D figure, we speak of its perimeter, area, and angles. The *perimeter* or the circumference of a 2-D figure is the distance around it. We use units of length to measure the perimeter. *Area* is the amount of coverage, or surface enclosed within the perimeter. We use square units to cover or to calculate the area. *Angles* are most commonly measured in daily life using *degrees*, whereas *radians* are useful units in more technical studies and applications.

To calculate the <u>perimeter</u> of 2-D shapes, we simply measure the distance around.

Here are some common formulas which make such measurements more efficient.

The perimeter of a square is **4 s**, where s is the length of a side.

The perimeter of a rectangle is **2a+2b**, where a and b are length and width.

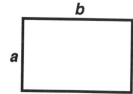

The perimeter of a triangle is **a+b+c**, where a, b, and c are the lengths of the sides.

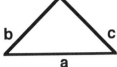

The perimeter of a trapezoid is **a+b+c+d**, where a, b, c, and d are the lengths of the sides

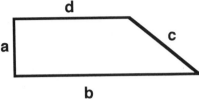

The perimeter of a parallelogram is **2a+2b**, where a and b are the lengths of the sides.

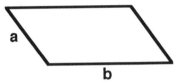

The perimeter of a circle is **π x D**, where D is the diameter of the circle, or **2πr**, where r is the radius of the circle.

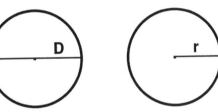

To calculate the <u>areas</u> of 2-D figures, the following formulas are helpful.

The area of a square is **s^2**, where s is the length of a side.

The area of a rectangle is **a x b**, where a and b are the lengths of the sides.

The area of a triangle is **1/2(b x h)**, where b is the length of the base and h is the altitude of the triangle.

The area of a circle is $\pi \times r^2$, where r is the radius of the circle.

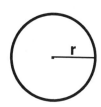

The area of a trapezoid is $1/2\,(b_1 + b_2) \times h$, where b_1 and b_2 are the lengths of the sides and h is the altitude.

The area of a parallelogram is $b \times h$, where b is the base and and h is the height or altitude.

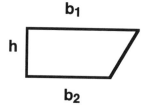

The Pythagorean theorem

Ancient Egyptians, thousands of years before Pythagoras was born, used his famous right triangle relationship in surveying their land following the flooding of the Nile. Mathematicians, however, have given credit to Pythagoras, and the relationship is known as the *Pythagorean theorem or property*. It states that in a right triangle the sum of the squares of the legs (a and b) is equal to the square of the hypotenuse (c). The resulting formula is $a^2 + b^2 = c^2$. The Pythagorean property is very useful in dealing with measurement involving right triangles.

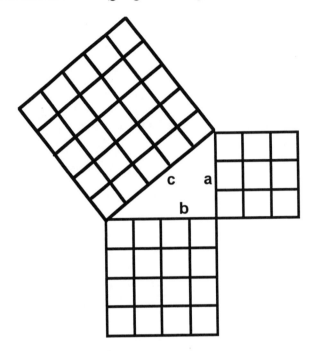

The classic diagram above illustrates this relationship. If the lengths of a = 3, b = 4, and c = 5, then the areas of the squares having sides a, b, and c, respectively, are 9, 16, and 25. The sum of the squares of a and b equals the square of c: $3^2 + 4^2 = 5^2$, or $9 + 16 = 25$. Thus, $a^2 + b^2 = c^2$.

When are two figures similar?

When two shapes fit exactly on top of each other, we say they are *congruent*. Two figures are *similar* if they have the same shape and all of their corresponding lengths are *proportional*. All squares are similar to each other, as are circles and regular polygons of the same number of sides.

Some properties of quadrilaterals (4-sided figures)

When connecting the midpoints of the sides of quadrilaterals, an internal parallelogram is formed, as shown in the following diagrams. There is an interesting relationship between the area of this parallelogram and the area of the original figure; that is, the area of the internal parallelogram is 1/2 the area of the original figure.

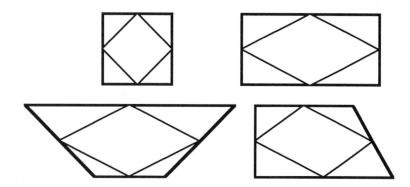

Symmetry of 2-D figures

A line that divides a figure into two congruent halves is called a *line of symmetry*. Some objects have only one line of symmetry (a). Others have two or more (b). Some objects are not symmetrical (are asymmetrical) (c).

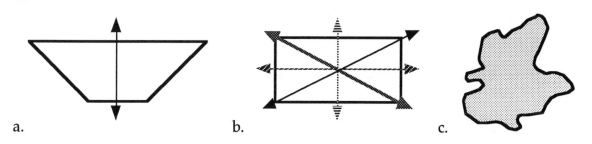

a.   b.   c.

C. NCTM position

K-4

Students will be able to describe, draw, and classify shapes.

Students will investigate the results of combining and changing geometric shapes.

Students recognize appropriate geometry in their world.

5-8

In addition to the above expectations, the students will:

Explore transformations of geometric figures.

Represent and solve problems using geometric models.

In addition to the above, 9-12 students should:

Apply properties of geometric figures.

Classify figures in terms of congruence and similarity and apply these relationships.

D. Integrating problem solving, reasoning, communicating, and making connections

Effective questioning is an essential part of a constructivist approach to teaching and learning. Accordingly, all the following activities begin with a question or set of questions. Many of the questions are divergent, allowing students to use critical thinking and creativity. Throughout the activities, students are encouraged to plan strategies, reason, use a variety of ways to communicate their ideas, and make meaningful connections as they are challenged to look for examples of 2-D geometry in a variety of settings.

E.   Prerequisite skills and knowledge

Students should have the skills and knowledge of problem solving, measuring, and appropriate units.

F.   Students' difficulties, confusion, and misconceptions

Many students have difficulty selecting appropriate units for the area of 2-D figures.

Many students confuse measurement of length with that for area.

Many students have difficulty estimating areas and perimeters.

A large number of students struggle with metric units.

Many children and adults believe if the shape of a figure is changed, the area will stay the same.

G. Factors contributing to students' difficulties, confusion, and misconceptions

The timing of when geometry is taught is inappropriate for many students. This is especially true in the case of formal geometry for high school freshmen or sophomores.

Inappropriateness of <u>how</u> we teach geometry contributes to lack of understanding.

H. Appropriate teaching strategies

The conceptual change strategy may be used with activities 1, 8, 9, 10, 11, and 12. Activities 2, 3, 5, 6, and 7 lend themselves to collaborative learning. Discussion and demonstration may be used with activities 4 and 5. The students could be encouraged to use visualization and mental model building also in activities 10, 12, and 13.

I. Teaching notes

The activities in this chapter cover a wide range. Some are appropriate for elementary school levels, while others are aimed at the secondary level. Many of the activities are designed to help students to come to make sense of the concepts by being involved and by DOING.

J.   Materials needed

graph or grid paper
magic markers
large sheets of paper
masking tape
Geoboards and rubber bands
calculators
measuring tapes
meter sticks
protractors

K.   Activities

See the following pages.

1. How many tiles? How much paint?
How much time to clean the windows?

Estimate each of the following situations, design a plan to check your estimate, then do it.

a.   How many tiles would you need to COVER the floor of the classroom?

b.   How many gallons of paint would you need to paint this classroom?

c.   How much time would it take to clean the windows in this classroom?

2. Who uses 2-D geometry?

As a group, bring examples from sports, business, school, and home where 2-D geometry is used.

3. Developing formulas for rectangle, square, triangle, trapezoid, circle, and ellipse

In your group, develop plans to to come up with the area of each figure, with the four sets of measurements that are given for each shape on the following table.

Kind of figure	Dimensions represented	If it is ...	The area is ?	If it is ...	The area is ?	If it is ...	The area is ?	If it is ...	The area is ?
rectangle	base & height	2 x 3		3 x 5		4 x 6		a x b	
square	side	3		5		8		x	
triangle	base & altitude	2 x 3		4 x 5		6 x 8		a x b	
parallelogram	base & altitude	4 x 6		6 x 8		10 x 12		a x b	
circle	diameter	6		10		12		d	

Based on your calculations, write the general expression for each shape.

4. Calculating areas of things that are difficult to calculate

Suppose that you only knew the RULE to calculate the area and the perimeter of a rectangle. Using appropriate materials, how would you use this knowledge to calculate each of the following areas:

a.   area of a triangle?

b.   area of a circle?

c.   area of a trapezoid?

d.   area of an ellipse?

5. Making conversions

What are the missing values for the following conversions?

a.   How many $cm^2 = 2m^2$?

b.   How many $dm^2 = 4m^2$?

c.   How many $in^2 = 3ft^2$?

d.   How many $ft^2 = mi^2$?

e.   How many $in^2 = m^2$?

How would you check your results?

6. Creating your own standard

By now you are familiar with the standard units of measurement for area, such as cm^2, m^2, in^2, ft^2. In your group, design a unit of measurement. Measure a few items and share your results with others. As a class, vote on the best unit, then explain why the vote went as it did.

7. Making a scale drawing

Using the ratios that are given, make a scale drawing of each of the following things:

a.   Top of teacher's desk   1/5

b.   Window   1/4

c.   Door   1/3

d.   Classroom floor   1/10

Begin by designing a plan.
What operations would you use, and why?

8. Using the Pythagorean property

Predict what would be the length of the hypotenuse for a right triangle with legs that measure 3 and 4.

How would you test your prediction?

What would be the hypotenuse of a right triangle with legs 6 and 8?

Do you notice a relationship between the dimensions of the base and height and those of the hypotenuse for a right triangle?

Test your hypothesis.

What professionals make use of this property?

9. Quadrilaterals (4-sided figures)

Predict what shape(s) would result if you connected the midpoints of the sides of each of the following quadrilaterals.

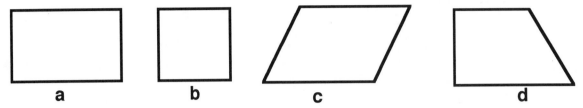

Share your predictions with others in your group.

Test your ideas.

Predict the relationship between the area of the resulting figure(s) and the area of the original shapes.

Share your predictions with others.

Test your ideas.

What general statement can be made about the shape(s) resulting from these midpoint connections?

10. Symmetry

Predict how many lines of symmetry are possible for each of the shapes below. Give reasons for your predictions.

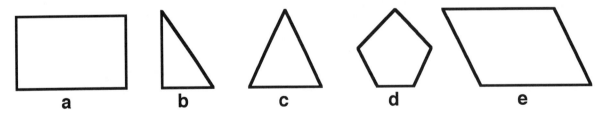

Share your ideas with others.

How would you test your ideas?

Get the necessary materials and test your predictions.

Is there a relationship between the number of sides of a regular polygon and the number of lines of symmetry? If so, what is the relationship? How would you test your idea?

11. Finding the pattern: number of angles, sides, degrees

A triangle--whether it is a right triangle, obtuse triangle, or equilateral triangle--has three sides with the sum of the interior angles equal to 180 degrees.

What is your prediction of the sum of the internal angles (number of degrees) for any 4-sided polygon?

Share your prediction with others.

How would you test your prediction?

What about a pentagon ? a hexagon? a decagon?

What general statement can you make about the relationship between the number of sides of a polygon and the sum of the degrees of the interior angles?

12. Similar or not?

Given the following sets of 2-D figures, predict whether or not the members of each set are similar.

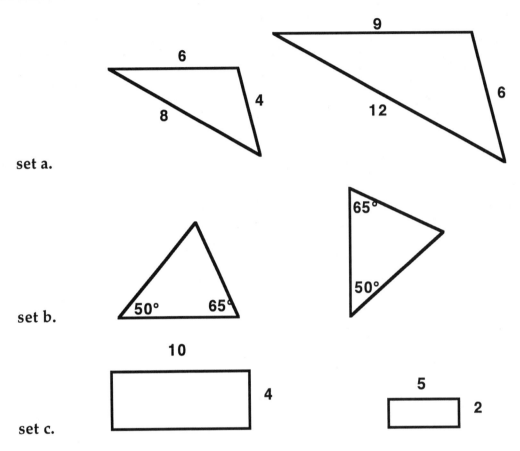

set a.

set b.

set c.

Share your ideas with others in the group.

How would you test your ideas?

Make a plan, then try it.

What did you find out?

What is needed for two figures to be similar?

13. Making congruent figures

Using the Geoboard, construct as many figures that are congruent to each of the following as you can. The figures may overlap.

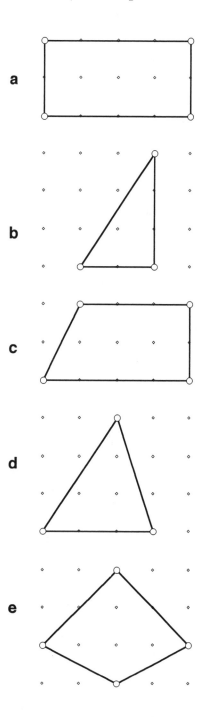

a

b

c

d

e

L.   Assessment ideas

Pencil and paper tasks and interviews of individual students or groups of students can be used to assess students' knowledge of 2-D geometry, specifically such aspects as area and perimeter, units, and conversions.

Performance assessment and students' projects can be used to assess students' ability to solve problems, make constructions, and use appropriate tools.

M.   Resources and references

Center for Occupational Research and Development. (1988). *Working with shapes in two dimensions.* Waco, TX: Author.

Elementary Science Study: Pattern Blocks. (1986). Nashua, NH: Delta Education.

Elementary Science Study: Mirror Cards. (1986). Nashua, NH: Delta Education.

Elementary Science Study: Geo Blocks. (1986). Nashua, NH: Delta Education.

National Council of Teachers of Mathematics. (1989). *Curriculum and Evaluation Standards for School Mathematics.* Reston VA: Author.

Neatrour, C. (1991). A strategy for discovering the formula for finding the area of polygons. *School Science and Mathematics, 91*(8), 362-366.

Pereira-Mendoza, L. (1993). What is a quadrilateral? *The Mathematics Teachers, 86*(9) , 744-776.

Reys, R. E., Suydam, M. N., & Lindquist, M. M. (1989). *Helping Children Learn Mathematics.* Englewood Cliffs, NJ: Prentice Hall.

Sobel, M. A., & Maletsky, E. M. (1975). *Teaching Mathematics: a Source Book of Aids, Activities, and Strategies.* Englewood Cliffs, NJ: Prentice Hall.

Toumasis, C. (1994). When is a quadrilateral a parallelogram? *Mathematics Teacher, 87,* 208-211.

7

FRACTIONS AND DECIMALS

A. Overview

This chapter includes definitions, critical concepts, language, and terms; interpretations, comparisons, operations, and applications of fractions and decimals.

B. Background information for the teacher

Many adults do not understand fractions or money and yet we continue to teach fractions as if they are a set of procedures. This chapter will emphasize developing a functional knowledge of fractions and decimals.

What are fractions?

Fractions are numbers that show how many parts of a whole or set are present. If I have 1/4 of a chocolate bar, I have one of four parts that make up the whole bar.

Several concepts are key to understanding fractions. The first two concepts are simple, but fundamental. First, the parts of the whole or the set must be equal in size. Nobody can have the "bigger half" of something. Second, you must be talking about the same-sized whole to do an operation with two units. For example, to add fractional parts of two pizzas to find out how much remains in relation to a whole pizza, both pizzas have to be the same size--the pizza slice isn't the same size for a large and a small pizza. These concepts are often glossed over and create misconceptions for children with different real-life experiences.

The *numerator* (top number) tells how many parts have been selected and the *denominator* (bottom number) represents the number of objects the whole or set contains in its complete form.

How are fractions interpreted?

Fractions can be interpreted in at least these ways: part of a whole, part of a set, a decimal, a ratio, an indicated division, an operator, and a measure. Fractions as an operator and as a measure are advanced concepts and are beyond the scope of this book.

The *part of a whole* interpretation is one in which the whole is divided into equal-sized pieces. For example, a sheet cake or candy bar has pieces that form the whole cake or candy bar. The fraction 3/5 represents a whole that is cut into 5 equal pieces, and we have 3 of those pieces. The *part-whole* interpretation is the least complex, and could be introduced in the early grades.

The *part of a set* interpretation involves individual pieces that are collected into a set. For instance, if two bicycles in a group are red and three are green, what part of the set of bicycles is green? 3/5. The bicycles are discrete objects that can combine to form a set of five objects, the denominator. The *part of a set* interpretation should be introduced after the *part-whole* interpretation so students do not confuse the two ideas.

A *decimal* is a fraction where the denominator is a power of 10. For example, 3/5 would have to be converted, finding an equivalent fraction to give it a denominator that is a power of ten (6/10 or 60/100). The standard algorithm is to divide the denominator into the numerator, which has the effect of converting it to a decimal (0.6 or six tenths). This means that we have six shares of a whole that has been divided into ten shares.

A *ratio* is another type of fraction interpretation. In a ratio, 3/5 would represent 3 out of 5 possible or 3 to 5 (3:5). Ratios convey the notion of relative magnitude. A *proportion* is a statement that equates two ratios. For instance, 3/5 = 6/10 is a proportion that demonstrates that 3 is to 5 as 6 is to 10.

The *indicated division* interpretation tells the student to do the operation. For example, 3/5 might be used to tell someone to divide 3 by 5. This procedure happens in middle school grades and is important for developing an understanding of rational expressions.

C. NCTM position

K-4

Increased attention to the meaning of fractions and decimals at this age level is recommended.

Children should be encouraged to understand fractions and decimals and not be expected to perform rote practice of paper-and-pencil calculations.

5-8

Students will develop operation sense applied to fractions and decimals and explore the relationships between representations of numbers during these grades.

Fractions and decimals should be integrated into a student's conception of number.

Decreased attention should be paid to memorizing rules and algorithms for fractions and decimals.

Exact forms of answers should not prevail over stressing the importance of using estimation for fractions and decimals as well as integers.

9-12

Arithmetic computation is assumed not to be a direct object of study in grades 9-12.

Instead, number and operation sense, estimation, and using these skills in problem solving and applications are stressed.

D. Integrating problem solving, reasoning, communicating, and making connections

Fractions and decimals provide excellent problem-solving situations, and comparing fractions and decimals involves numeric reasoning. Effective communication requires an understanding of fractions and decimals. Fractions and decimals are found in other disciplines such as science, social studies, sports, and music, as well as many aspects of daily life.

E. Prerequisite skills and knowledge

Students should have an excellent understanding of whole number operations before moving into operations involving fractions and decimals. Place value concepts should be in place before teaching decimals. Students need to understand the concept of *equal* so that equal-sized pieces can be understood.

F. Students' difficulties, confusion, and misconceptions

Students' misconceptions about fractions and decimals are based on a lack of understanding of the meaning of fractions. Problems arise especially when multiplying and dividing fractions.

For instance, when two whole numbers are multiplied or added the result is larger. Many students translate this to fractions and assume that whenever two fractions are multiplied, the result should be larger.

Conversely, division of whole numbers yields a smaller result. With fractions, the unit changes to what you are dividing by and the answer can be larger than either fraction.

G. Factors contributing to students' difficulties, confusion, and misconceptions

When students do not understand whole-number operations in a complete sense, operations involving fraction concepts can be mangled. Multiplying numbers between zero and one is one instance.

In the case of $\frac{1}{3} \times \frac{1}{2} = \frac{1}{6}$ it is difficult for many students to believe that multiplication can make something smaller. Their difficulty comes from the belief that multiplication is <u>only</u> like doing a lot of addition and addition always makes something larger.

Another case is division. When $\frac{1}{3} \div \frac{1}{2} = \frac{2}{3}$, many students cannot get beyond the rule of "invert and multiply" to realize the question of how many (or what part of the) one-halves fit into one-third. The result is two-thirds of a one-half piece. For example, consider the following diagrams.

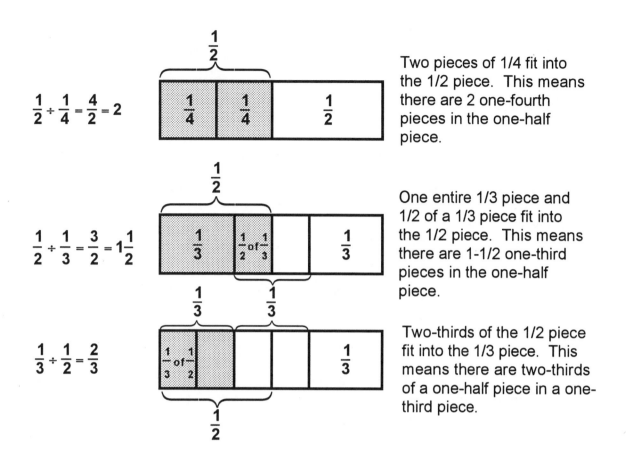

$$\frac{1}{2} \div \frac{1}{4} = \frac{4}{2} = 2$$

Two pieces of 1/4 fit into the 1/2 piece. This means there are 2 one-fourth pieces in the one-half piece.

$$\frac{1}{2} \div \frac{1}{3} = \frac{3}{2} = 1\frac{1}{2}$$

One entire 1/3 piece and 1/2 of a 1/3 piece fit into the 1/2 piece. This means there are 1-1/2 one-third pieces in the one-half piece.

$$\frac{1}{3} \div \frac{1}{2} = \frac{2}{3}$$

Two-thirds of the 1/2 piece fit into the 1/3 piece. This means there are two-thirds of a one-half piece in a one-third piece.

H. Appropriate teaching strategies

Mental model building, teaching for conceptual change, and discussion and demonstration are useful strategies for helping students develop concepts related to fractions and decimals.

Mental model building is important in constructing an understanding of fractions. Students should use manipulatives in diverse ways so they have to opportunity to think about fractions as more than just pieces of pie or pizza. The circle is not an easy model for children to divide into thirds without an understanding of a circle equaling 360°. The rectangle of a fraction bar or paper folding, the hexagon of pattern blocks, and the rectangular prism of Cuisenaire rods all provide a variety of models to develop understanding of fractions. Power-of-ten blocks are useful 3-dimensional models for decimals, and decimal squares provide a 2-dimensional model. Activity 5 asks children to evaluate their mental models so they become able to critique and improve upon models.

The conceptual change strategy is used in activity 3, which is used to challenge students to think about operations with fractions so they can formally construct their ideas. Discussion and demonstration encourage students to explain why operations with fractions work as they do. It is important that the next generation of learners understands why to invert and multiply when dividing with fractions, as well as to discuss what makes a real-life fraction situation, as in activity 4.

I.   Teaching notes

It is important to establish that units are the same size <u>before</u> a common denominator is found. When checking student diagrams, therefore, make sure the wholes are the same size. For example, the fractions below would be difficult to compare because the wholes are not the same size. When comparing fractions, it is assumed the wholes are the same size.

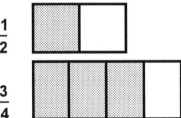

During the number line activity, ensure that students think about the operations and the type of fractions they are dealing with. You might need to ask probing questions to make sure they get as much out of the activity as possible.

J.   Materials

> fraction bars
> fraction manipulatives such as:
>> Cuisenaire rods
>> pattern blocks
>> fraction squares
>> paper for folding

K.   Activities

See the following pages.

1. Comparing fractions

Predict which fraction is greater than the other. Then draw a diagram that demonstrates whether each fraction is larger, smaller, or the same as the other fraction.

Example:

The situation: **Compare** $\frac{1}{2}$ **and** $\frac{2}{4}$

Prediction: I think that ....

Making the comparison:

$\frac{1}{2}$

$\frac{2}{4}$

Conclusion: $\frac{1}{2}$ **is equal to** $\frac{2}{4}$.

Make your own diagrams to compare the following sets of fractions.

a. $\frac{4}{10}$ and $\frac{2}{5}$

b. $\frac{2}{5}$ and $\frac{4}{8}$

c. $\frac{5}{8}$ and $\frac{3}{5}$

d. $\frac{1}{5}$ and $\frac{2}{10}$

e. $\frac{3}{4}$ and $\frac{5}{6}$

f. $\frac{1}{6}$ and $\frac{2}{3}$

g. $\frac{7}{4}$ and $\frac{4}{7}$

h. $\frac{1}{2}$ and $\frac{1}{3}$

i. $\dfrac{6}{4}$ and $\dfrac{7}{5}$ j. $\dfrac{3}{9}$ and $\dfrac{1}{3}$

k. $\dfrac{3}{8}$ and $\dfrac{1}{4}$

What patterns do you see that will help make the comparison of two fractions easier?

What role does the denominator play in your patterns?

What role does the numerator play in your patterns?

Can you prove your ideas about comparing two fractions to your group?

As a group, agree on and write a general statement about how to compare two fractions to tell if one is bigger than the other.

2. Ordering fractions and decimals

Using your rule for ordering fractions, order the following fractions and decimals from smallest to largest.

$$\dfrac{2}{5}, \quad .5, \quad \dfrac{6}{7}, \quad .27, \quad \dfrac{7}{8}, \quad .93, \quad .44$$

What process did you use to put these fractions and decimals in order?

Does it fit with your earlier rules for fractions?

What do you think you would have to do to order the fractions below?

$$.49, \quad \dfrac{5}{2}, \quad 2.5, \quad \dfrac{7}{5}, \quad 3.5, \quad \dfrac{2}{3}, \quad \dfrac{3}{2}, \quad .75, \quad 1.25, \quad \dfrac{5}{4}$$

What is the order of these numbers from least to greatest?

What process did you use to figure this out?

How did this compare to your earlier process?

3. Using the number line to think about operations with fractions

Set A.

Use the number line to decide if the answer to each of the fraction problems shown below would be in region a, b, or c of the number line.

Situation a. $\dfrac{3}{4} + \dfrac{2}{3}$

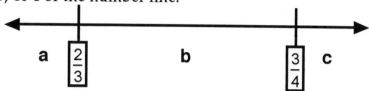

Situation b. $\dfrac{3}{4} - \dfrac{2}{3}$

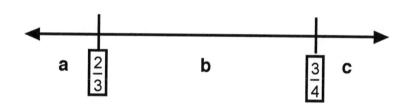

Situation c. $\dfrac{3}{4} \times \dfrac{2}{3}$

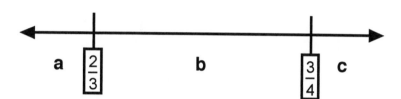

Situation d. $\dfrac{3}{4} \div \dfrac{2}{3}$

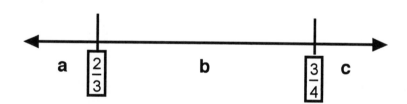

Based on these examples, what pattern(s) do you observe that could be used to solve other problems involving fractions?

Set B.

Next, use the pattern(s) you observed in the first set of situations to solve the following problems.

Situation a. $\dfrac{5}{4} + \dfrac{4}{3}$

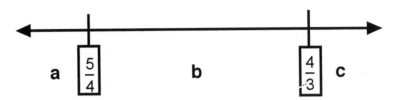

Situation b. $\dfrac{5}{4} - \dfrac{4}{3}$

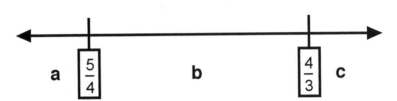

Situation c. $\dfrac{5}{4} \times \dfrac{4}{3}$

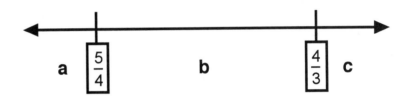

Situation d. $\dfrac{5}{4} \div \dfrac{4}{3}$

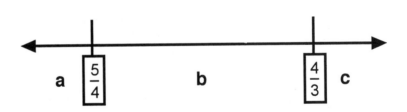

Did your original pattern(s) work for these problems? Why or why not?

What new pattern(s) for predicting do you observe, if any?

Can you find any cases where your new pattern(s) might not work?

C.

Use number lines to show operations involving other, different fractions.

4. Creating your own fraction and decimal problems

Create your own real-life word problem for each of the problems a through i below.

a. $\dfrac{2}{5} + \dfrac{3}{4} =$

b. $\dfrac{3}{5} - \dfrac{1}{5} =$

c. $\dfrac{3}{5} - \dfrac{1}{4} =$

d. $\dfrac{2}{3} \times \dfrac{1}{2} =$

e. $\dfrac{1}{5} \times \dfrac{1}{5} =$

f. $\dfrac{3}{5} \div \dfrac{1}{5} =$

g. $\dfrac{1}{4} \div \dfrac{1}{2} =$

h. $\dfrac{5}{3} \div \dfrac{1}{3} =$

i. $\dfrac{5}{3} \times \dfrac{1}{2} =$

Give your word problems to your neighbor to solve.

What can you say about the answer to a division problem that involves fractions between zero and one?

What can you say about the answer to a multiplication problem that involves fractions between zero and one?

How does what you wrote in response to the previous two questions change for fractions that are larger than one?

This same activity could be done with decimals.

5. Representing fractions and decimals --what makes a good model?

Using one of the word problems your group created in activity 4, draw a diagram that best represents what is happening in the problem.

Have each student in your group draw a diagram that best represents what is happening in the first problem in activity 4, which is 2/5 + 3/4.

Compare your representations and discuss which diagram the group thinks best represents the problem and explain the reasons.

What makes a representation better than another to show what mathematics is going on in the problem?

Continue, choosing remaining problems from Activity 4 and diagram the situations. Compare representations with your group to find good diagrams for all of the problems. Decimals can be interchanged with fractions in this activity and the same concepts should be evident.

Compare your findings of what makes a good representation or diagram with the rest of the class. Demonstrate your findings.

L. Assessment ideas

Portfolios and journals allow students to demonstrate understanding or misunderstanding through the choices they make about what to include and write about. Look for the diagrams of fractions that students decided were the best representations of the concepts involved as well as word problems students write to match given problems to evaluate their concepts of fractions and decimals.

Ask students to write their own problems with the teacher providing the operation sentence and the students writing the word problem and solution strategy. These problems can reveal what operations and algorithms the students understand.

Have students write their own word problems in class and give them to other students to solve and evaluate, then discuss in the small groups and in the class as a whole.

M. Resources and references

Behr, M. J., Harel, G., Post, T., and Lesh, R. (1992). Rational number, ratio, and proportion. In D. A. Grouws (Ed.), *Handbook of Research on Mathematics Teaching and Learning*, 296-333. New York: MacMillan.

Behr, M. J. & Post, T. R. (1988). Teaching rational number and decimal concepts. In T. R. Post Ed.) *Teaching Mathematics in Grades K-8: Research Based Methods.* Boston: Allyn and Bacon.

Behr, M. J., Wachsmuth, I., Post, R. , and Lesh, R. (1984). Order and equivalence of rational numbers. *Journal for Research in Mathematics Education, 15,* 323-341.

Mack, N. K. (1990). Learning fractions with understanding: Building on informal knowledge. *Journal for Research in Mathematics Education, 21,* 16-32.

National Council of Teachers of Mathematics. (1989). *Curriculum and Evaluation Standards for School Mathematics.* Reston VA: Author.

Post, T. R. (1992). *Teaching Mathematics in Grades K-8: Research-Based Methods.* Boston: Allyn and Bacon.

8
THREE-DIMENSIONAL GEOMETRY

A. Overview

Concepts included in this chapter are: examples of 3-D shapes, relationship between two- and three-dimensional shapes, surface area and volume of 3-D shapes, and applications of three-dimensional geometry.

B. Background information for the teacher

As the names implies, 3-dimensional shapes have three dimensions--length, width, and height. Two-dimensional shapes can be <u>covered</u>, but cannot be <u>filled</u>. Three-dimensional shapes can <u>hold</u> something, or they are <u>solid</u>, meaning they occupy space. This characteristic is possible because these shapes have a third dimension, sometimes called depth and sometimes called height. Some common 3-D shapes are rectangular prisms, cubes, cylinders, spheres, hemispheres, cones, and pyramids.

What is the relationship between 2-D and 3-D figures?

If we project the shadow of a sphere on the wall, we only see a circle. If we project the shadow of a cone laterally on the wall we see a triangle. If we project the shadow of a rectangular prism laterally on the wall, we see a rectangle. Imagine projecting a cylinder. We might see a rectangle or a circle, depending on how we orient the cylinder. If we project the shadow of a cube, we see a square. Three-D shapes have much in common with 2-D shapes, but they have depth. Projection on the wall does not show depth.

Terms and units used in 3-D geometry

We describe 3-D objects by their shape, surface area, and volume. *Surface area* is what can be covered. When talking about surface area, it is essential to be clear about whether we are referring to *total surface area* (the external surface of the entire figure) or *lateral surface area* (the surface area excluding the top and bottom of the solid). *Volume* is the capacity of the object, how much room it takes or how much it can hold.

In the CGS system, the unit for surface area is the same as that for area of 2-D figures; for example, cm^2. The unit for volume is cm^3; however, when we speak of fluids (liquids and gases), we normally use milliliter (ml) as the unit for volume. One ml is the same as $1\ cm^3$.

In the MKS system, the unit for surface area is m^2 and the unit for volume is m^3. For fluids (liquids or gases), we use liters.

In the United States, we typically use in^2 and ft^2 to measure surface area of 3-D figures. In^3, ft^3, quarts, and gallons are used for volume measurement. For example, we may measure the amount of gas in a balloon in in^3 and the amount of milk in quarts.

Some common conversions associated with 3-D figures are:

4 quarts = 1 gallon.
1 gallon = 231 in^3.
$10^6\ cm^3$ = 1 m^3.

What are some examples of 3-D shapes in our environment?

An athletic facility is a place where you will see various examples of 3-D shapes. Basketballs, volleyballs, soccer balls, ping pong balls, and tennis balls are spheres. Mats are rectangular prisms (boxes). Relay batons are cylinders. Conical boundary markers may be used to mark corners or lanes.

In the classroom we see cabinets and chalkboard erasers in the shapes of rectangular solids. Chalk sticks, laboratory beakers, and graduated cylinders are cylinders. An Erlenmeyer flask is a cone with a cylinder for its top.

Houses are usually built in shapes involving right angles (rectangular solids or cubes), but geodesic dome homes are almost spherical. A-frame homes and many contemporary designs with unusual angles create diverse 3-D shapes and spaces. Lumber for home-building may be in the form of rectangular solids (such as 2 x 4s) or cylindrical logs. Swimming pools also are built in a diversity of symmetrical and asymmetrical shapes.

Storage silos on farms and ranches are usually cylinders. Bales of hay may be rectangular solids or cylinders.

Food producers use a variety of 3-D shapes for packaging food, such as cylinders for canned foods and beverages and cubes and rectangular solids for cereals and crackers. A milk carton is a combination of a rectangular solid and a triangular prism.

Artists and crafts people work in a variety of 3-D shapes, for example, when they are sculpting, creating ceramics, making jewelry, or blowing glass. Even a painter, designer, or architect working in a two-dimensional medium may depict three-dimensional shapes. Computer graphics, including those used in computer games and special effects in motion pictures, create successful illusions of 3-D shapes.

A mechanic uses various 3-D shaped tools and parts. Overall, 3-D shapes are everywhere in our constructed environment, ready to be noticed.

Three-dimensional shapes are also found in nature. What is the basic shape of the Moon? the Earth? What about other planets in our solar system? What do you think is the reason for their shapes? What shape is a fir tree? a salt crystal? a crystal of quartz or amethyst? an icicle?

How do we calculate the surface area and volume of different 3-D shapes?

Mathematical formulas have been derived for using measurements to calculate surface area and volume. A challenge in teaching and learning, however, is to start with the actual, physical shape to develop a concrete concept before applying the mathematical abstraction in order to minimize confusion and misconceptions for the students.

Cylinder

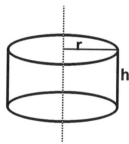

Lateral area = 2 •π• r • h --or-- $2\pi r h$

Total surface area = (2 •π• r2) + (2 •π• r • h) --or-- $2\pi r^2 + 2\pi r h$

Volume = π• r2 • h --or-- $\pi r^2 h$

Cube

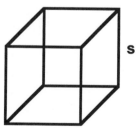

Total surface area = 6 • s2

Volume = s3

Rectangular prism

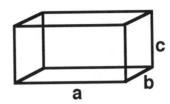

Total surface area = 2(a • b) + 2(b • c) + 2(a • c)

Volume = a • b • c --or-- abc

Cone

Lateral area = L • π • r --or-- Lπr

Total surface area = (πr2) + (Lπr)

Volume = $\frac{1}{3}$ π r^2 • H

Sphere

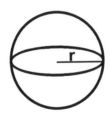

Area = 4 πr2

Volume = $\frac{4}{3}$ π r^3

C. NCTM position

K-4

Students will develop a spatial sense

Students will recognize and appreciate 3-D geometry in their world

5-8

In addition to the above accomplishments, students will:

Identify and describe 3-D figures.

Visualize and represent geometric figures.

Represent and solve problems using geometric models.

Develop a sense to distinguish between 2-D and 3-D figures.

Understand and apply properties and relationships involved in 3-D geometry.

Develop an appreciation of geometry as a means of describing the physical world.

9-12

In addition to the above accomplishments, the students will:

Interpret and draw 3-D objects.

Apply properties of 3-D geometry.

Classify figures in terms of congruence and similarity and apply these relationships.

D. Integrating problem solving, reasoning, communicating, and making connections

Use of effective questioning with students is essential to guide appropriate understanding of 3-D geometry. As students are involved in various activities, projects, and discussions, they should be encouraged to make use of problem solving skills, reasoning, and communicating about 3-D figures.

E. Prerequisite skills and knowledge

For certain activities, students will need prior knowledge of 2-D geometry, measurement, angles, and units, as well as the ability to visualize.

F. Students' difficulties, confusion, and misconceptions

Many students have difficulty visualizing 3-D figures from 2-D blueprints.

Many students have difficulty using appropriate units for 3-D geometry.

Many students have difficulty associating 2-D & 3-D figures.

Many students intuitively believe that volumes created by the same amount of material are necessarily the same.

G. Factors contributing to students difficulties, confusion, and misconceptions

Traditional teaching of geometry progresses from point to line to plane and then on to three dimensions, which is opposite to the sequence in which most learners learn and to the way in which they experience the world.

Instruction emphasizes terms and formulas before the learner is ready.

There is a mismatch between how a learner learns and develops and the assumptions the instructional materials and teaching strategies make.

Most geometric concepts are abstract and require formal thinking, which many students do not yet have.

H. Appropriate teaching strategies

Among others, strategies appropriate for teaching 3-D geometry are teaching for conceptual change, mental model building, discussion and demonstration. We suggest using the teaching for conceptual change strategy with activities 1, 3, and 5. In activities 11 and 12, students use mental model building. Collaborative learning may best be used with activities 2, 4, 14, and 15. Activities 6, 7, 9, and 10 lend themselves to classroom discussion.

I. Teaching notes

Most of the activities in this section call for hands-on manipulation. This emphasis does not minimize the importance of formal geometry, but does focus on the need for relevant concrete experiences as a basis for understanding formal geometry. We have targeted the DOING of mathematics through the geometry of 3-D figures. As

have targeted the DOING of mathematics through the geometry of 3-D figures. As with other chapters, our intent is not to provide recipes and remedies to deal with every student's misconception and point of confusion related to 3-D geometry. We believe the activities and teaching strategies proposed here will help students to recognize and deal with their confusion and misconceptions.

On activities dealing with shadows of three-dimensional figures, you may use the overhead projector and project each side on the wall, while hiding the object behind a partition.

J. Materials

overhead projector
flashlight
graph or grid paper
cellophane tape
cm cubes
pattern blocks
Geoblocks
3-D figures: cubes, rectangular prisms, spheres, cylinders, cones, etc.
beverage cans, small coffee cans, large coffee cans
plastic or metallic rectangular solids
containers in the shape of a half sphere
rice, corn, or sand for filling cylinders
calculator
250-500 graduated cylinders
measuring tapes
rules and meter sticks
micrometers
calipers
thick but bendable wire or pipe cleaners
soap solution: 1 part liquid detergent, 3 parts water, 2.5 parts glycerin

K. Activities

See the following pages.

1. Relationship between 2- & 3-dimensional shapes

Shown the shadow of one side of a 3-D figure, can you guess what the shape is?

Without actually seeing the shape, what else would you need to know in order to make a better guess?

2. Units associated with 3-D shapes: surface area and volume

Using graph paper, cellophane tape, and cm cubes, construct the boxes described in the table. Then,

a.   count the number of <u>squares</u> that are <u>covering</u>
   the sides
   the entire box

b.   count the number of <u>cubes</u> that are needed to <u>fill</u> the box.

Side measure-ments	Number of squares to cover sides	Number of squares to cover entire box	Number of cubes to fill the box
Cubes: 3 cm			
5 cm			
Rectangular solids: 2 x 3 x 4			
3 x 5 x 6			

Calculate the same information as in the chart above for solids having the following dimensions:

         5 x 12 x 15

         X x Y x Z

3. Covering and filling 3-D shapes

Use an ordinary, 8-1/2 x 11 in sheet of paper. The paper can be rolled two different ways--lengthwise or widthwise--to construct a cylinder.

a.   Covering

In predicting how much of a material it would take to <u>cover</u> the outside of each cylinder, which of the following statements do you think is true?

It takes the same amount to cover the outside of both cylinders.

It takes more to cover the cylinder formed by rolling the paper lengthwise.

It takes more to cover the cylinder formed by rolling the paper widthwise.

Give reasons for your prediction.

How would you test your prediction and explanation?

b.   Filling

In predicting how much of something it would take to <u>fill</u> each of the cylinders you have constructed, which of the following statements do you think is true?

It takes the same amount to fill both cylinders.

It takes more to fill the cylinder formed by rolling the paper lengthwise.

It takes more to fill the cylinder formed by rolling the paper widthwise.

Share your prediction and explanation with others in your group.

Have someone from your group share the predictions and the explanations of individual members of the group with the rest of the class.

How would you test your ideas? Get the necessary materials and test your predictions.

After performing the tests, do you want to make any changes in your explanations?

As a small group, come up with a statement explaining how you can make sense of these observations.

What are some examples and applications of this phenomenon?

What are some other questions and problems related to what you observed that you would like to investigate?

c. Shape relationships

For the cylinders you have constructed, consider the following challenge.

If you cast shadows of the surface of each cylinder, what would be the shape of the shadow? How would the sizes of the shadows compare?

Share your predictions and explanations with others in your group.

Have someone share the predictions and the explanations of the members of your group with others in the class.

How would you test your ideas?

Get the necessary materials and test your predictions.

In your group, come up with a statement that makes sense of your observations.

What are some examples and applications of what you have observed here?

Think of additional questions and problems that you would like to pursue related to the geometry of shadows.

4. Comparing measurements and calculations of surface area and volume

You have access to the following items: a pop can, a small coffee can, a container in the shape of half a sphere, and a plastic or metallic rectangular solid.

In your group, design ways that you can (a) <u>measure</u>, and (b) <u>calculate</u> the surface area and the volume of each shape.

If there are discrepancies between your measurements and your calculations, how would you account for them?

Determine the accuracy of the claims made by the manufacturers about how much food or drink there is in a given container.

5. Making shapes from other shapes

There are many 3-D shapes which can be made from other 3-D shapes. Your challenge is to make some given shapes from other shapes <u>and</u> to find other examples not given here. Share what you have constructed with others.

a.   Make a cylinder from cones.

b.   What shapes can you use to make a drinking cup?

c.   Find other examples

How can you use what you observed in this activity to determine the volume of one shape from the volume of other shapes?

6. Examples of 3-D figures in nature, at home, and in school

Individually or with some partners, bring to class examples of 3-dimensional shapes you observe at home, in school, and in nature.

Provide a rationale for why you selected your examples.

7. Measuring things that are not easily measurable

In the classroom, in the gym or at home, we encounter variety of 3-D shapes. Bring to class two objects whose surface areas and volumes can be measured easily. Also bring two objects whose surface areas and volumes cannot be measured easily.

Calculate and the report the surface areas and volumes of the first set.

Design ways that you can calculate the surface areas and volumes of the items which are not easy to measure. How would you check your calculations?

8. Choosing appropriate tools and units to measure 3-D shapes

You will be provided with geometric objects of various shapes and sizes. Select appropriate units and instruments which you will use to measure the surface area and volume of each shape.

Explain the rationale for your choice of units and instruments.

9. Describing a 3-D shape

A member of the class has access to several 3-D shapes. The member will describe attributes of each object, in turn. Your challenge is to build a mental model of each shape that is described.

How did your model compare with the actual object?

10. Why do people use different 3-D shapes?

Contact various people, including scientists, engineers, contractors, farmers, bakers, machinists, sculptors, architects, furniture manufacturers, or others.

From at least five of these people solicit information as to their rationale for using some particular 3-D shapes.

11. Geoblocks

a.   Shown below are all the faces of one particular block drawn in outline--this is one object. Your challenge is to find the block that goes with this set. The numeral **2** means that the block looks the same on two of its sides.

 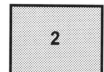

b.   .Shown several views of a block structure drawn in outline, build the structure.

c.   Present other members of your group with your own drawings of faces of a particular block. Challenge them to make a mental model of the block.

12. Which patterns make a rectangular prism, which make a cube?[3]

a.   Use grid paper to make boxes which are cubes and others which are rectangular solids.

Carefully undo what you have constructed and identify the patterns of construction by outlining each side of the structure.

How many different patterns can you construct that will make rectangular boxes and cubes?

b.   Look at patterns a-g, on the following page.

Which of the patterns will construct a cube?

Which of the patterns will construct a rectangular prism?

How did you reach your conclusions?

How could you test your conclusions?

---

[3] Adapted from Sobel, M. A. and Maletsky, E. M. (1975). *Teaching Mathematics: a Source Book of Aids, Activities, and Strategies*. Englewood Cliffs, NJ: Prentice-Hall.

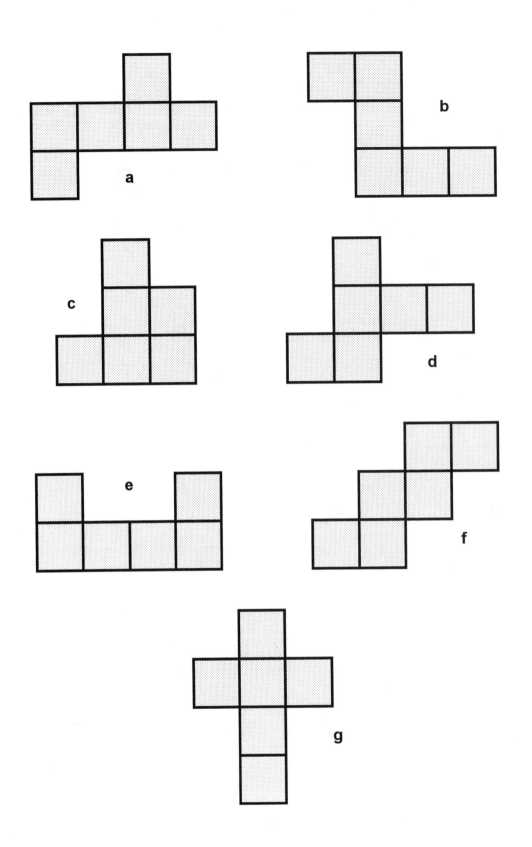

13. Relationship between surface area and volume

Study the formulas for the surface area and the volume of each of the following shapes: cylinder, sphere, rectangular solids, cone.

What statement can you make about the relationship between surface area and volume for different 3-D shapes?

14. We have built a structure that is 10 x 10 x 10

Here are some things we want to know about the structure:

a.   How many 1 x 1 x 1 cubes can we fit into it?

b.   How much material do we need to cover the entire structure?

c.   If we wanted to paint the sides of the structure, how much area would we have to paint?

If we change the structure by making it 10 x 5 x 20, would there be a change in b and c above?

How did you arrive at your answers?

How would you test your ideas?

15. Geometry of soap bubbles

In your group, build 3-dimensional frameworks like those pictured below, using thick wire or pipe cleaners.

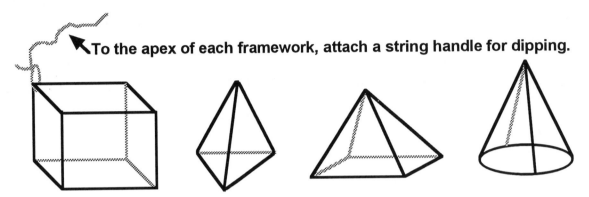

To the apex of each framework, attach a string handle for dipping.

Predict what shapes will result when each structure is dipped into the soap solution.

Give reasons for your predictions.

Share your predictions and explanations with others in your group.

Have someone from your group present to others the predictions and explanations of each member.

Get the necessary materials and test your ideas.

How do your observations compare with your predictions?

If there are disagreements between your predictions and your observations, what changes do you want to make in your explanations?

What generalizations can you make about your observations?

Share your statements with others in the class.

What are some applications of this activity?

What real-world examples can you think of?

What additional questions would you like to pursue related to the geometry of soap bubbles?

L. Assessment ideas

Pencil-and-paper assessment may be used to assess students' knowledge of geometric facts, including formulas, units, conversion of units, calculating surface area and volume of various 3-D shapes.

Student projects, such as those in activities 6, 7, 8, 9, and 10, may be assessed on their completeness, originality, and presentation.

Students' ability and willingness to make predictions, come up with explanations, and test and revise their explanations may be assessed by observing students.

Students' ability to visualize and construct mental models, such as in activities 11 and 12 may be assessed by having students present their mental models.

M. Resources and references

Center for Occupational Research and Development. (1988). *Working with Shapes in Two Dimensions.* Waco, TX: Author.

Elementary Science Study: Geo Blocks. (1986). Nashua, NH: Delta Education.

Grouws, D. A. (1992). *Handbook of research on Mathematics Teaching and Learning.* New York, NY: MacMillan.

Hopley, R. B. (1994). Nested platonic solids: A class project in solid geometry. *Mathematics Teacher, 87,* 312-318.

National Council of Teachers of Mathematics. (1989). *Curriculum and Evaluations Standards for School Mathematics.* Reston, VA: NCTM.

National Research Council. (1996). *Mathematics and Science Education Around the World: What Can We Learn from the Survey of Mathematics and Science Opportunities (SMSO) and the Third International Mathematics and Science Study (TIMSS)?* Washington, DC: National Academy Press.

Sobel, M. A., & Maletsky, E. M. (1975). *Teaching Mathematics: a Source Book of Aids, Activities, and Strategies.* Englewood Cliffs, NJ: Prentice Hall.

Sunburst Communications, (1987). *The Super Factory.* Pleasantville, NY: Sunburst Communications.

9
MEASUREMENT

A. Overview

Concepts included in this chapter are: importance of measurement, need for standards of measurement, units of measurement (CGS, MKS, English), estimating, fundamental units, derived units, measuring tools, accuracy, precision, and tolerance.

B. Background information for the teacher

Why should we pay attention to measurement?

Measurement is a part of everyday life and many jobs. Many disciplines rely on measurement. Measurement is an important process for collecting data and information in order to answer questions. A basic premise in measurement is that we need common reference points to measure things so there is a "fairness" in consumption. Children can begin by using arbitrary measures such as their own hands, feet, or other non-standards units. Teachers can point out the inadequacy of these measures by having students apply them to various items--the idea of fairness is important to young children. The next step to develop a concept of measurement is to use Unifix cubes, paper clips, or a more standard unit of measurement. Students then can use more formal standard measures, such as the metric system.

The units of measurement

When speaking of units, we should distinguish between fundamental units and derived units. As the name implies, *fundamental units* are based on a fundamental characteristic of nature. These are units of length, mass, and time.

In the metric system, these include: Meter, Kilogram, and Second, abbreviated, MKS. *Meter (M)* is a unit of length. A meter is equivalent to 1,650,763.73 wavelength of the orange-red light emitted by Krypton-86. *Second (S)* is a unit of time. It is equivalent to 9,192,631,770 vibrations of Cesium-133. *Kilogram (Kg)* is a unit of mass (the amount of material). It is based on the atomic mass of the Carbon 12 atom. One-twelfth of a Carbon-12 atom is called the *atomic mass unit (u)*. One u is equivalent to $1.6605655 \times 10^{-27}$ Kg. Also, one liter of water has a mass of one kilogram.

CGS is another set of units in the Metric System. They are abbreviations for *Centimeter*, *Gram*, and *Second*.

Foot is a unit of measurement. It is equal to 1/3 of a *yard* and to 12 inches. An *inch* is equal to approximately 2.54 cm. A story (difficult to believe) about how the yard came to be a unit of measurement is that a yard represented the length of the arm of King Henry I. If this were true, he must have been a giant, with arms 36 inches (91 cm.) long.

Units such as those for speed, energy, and temperature are *derived units*. They are derived from the fundamental units of length, mass, and time. For example the unit of speed, whether cm/sec or miles/hr is derived from the ratio of length to time.

Some examples of standard units

Length:	cm, m, in, foot
Area:	cm^2, m^2, in^2, ft^2
Volume:	cm^3, m^3, in^3, ft^3
Mass:	gm, kg, pound
Temperature:	F, C, K
Speed:	cm/sec, m/sec, miles/hr
Energy:	Joules, erg, BTU, calorie

Why do we need standard measurements?

One of the most important reasons to have standards for measurement is for communication. When we say this page is 25 cm long, we all know what is being said. This is because cm is a standard. It means the same thing to everyone familiar with this unit. If, however, we said the page is an arm length, would someone else know how long the page is? Whose arm are we using as standard? It is not clear.

Measurements and measuring tools

Whether cooking in our kitchen, building a deck, planning a trip, or engaging in nursing or an engineering profession, we rely a great deal on measurement.

We can estimate length, mass, temperature, and time to a certain extent. We have had to develop technological extensions of our senses, however--such as the thermometer, balance, telescope, microscope, and an array of data sensing and recording instruments--to help us collect more accurate data that can be quantified, compared, and communicated.

Suppose we want to measure the length of a room in our house, or we are working on a project where we need to measure the thickness of a sheet of paper. Do we use the same tools for both tasks? Given choices among a meter stick, micrometer, and caliper, we need to decide which tool is more appropriate to accomplish the task. With a meter stick, we can measure the length of a room to the closest centimeter, meeting our needs for that task; but a meter stick is not an appropriate tool to measure the thickness of a sheet of paper.

Accuracy, precision, and tolerance

Accuracy is an indication of how close the results of a measurement come to the true value. The measurement is not exact because neither the measuring process nor the instrument used is exact. *Precision* is a measure of how identically a measurement is repeated, without reference to to a "true" or "real" value. In a measuring instrument, precision is defined as the smallest fraction or decimal division on the instrument. *Tolerance* is the greatest range of variation that can be allowed in the dimensions of a manufactured part for it to be acceptable and fit together with other parts.

C. NCTM position

K-4

Students will understand the attributes of length, mass, area, volume, time, temperature, and angle.

Students will develop the process of measuring and using the units of measurement.

Students will make and use estimates of measurement.

Students will use measurement in everyday situations.

5-8

In addition to the above expectations, students will:

Use measurement to describe and compare phenomena.

Select appropriate tools to measure to the degree of accuracy required in a particular situation.

Understand the structure and use of systems of measurement.

Develop the concept of derived units.

Develop formulas and procedures for determining measures to solve problems.

D. Integrating problem solving, reasoning, communicating, and making connections

Measurement activities lend themselves to using problem solving, reasoning, communicating and making connections. Problem solving and reasoning are natural parts of this unit. As students convert units, graph, and use words, they are communicating about mathematics and in mathematical forms. There also are many opportunities for students to make connections. As they use measurements in various situations, they can connect mathematics to other disciplines.

E. Prerequisite skills

Students should be able to appropriately use the measurement tools. To convert units and make sense of the units for area and volume, they should be able to understand ratio and proportion.

F. Students' difficulties, confusion, and misconceptions

Many students find it difficult to understand the units for area and volume; *e.g.,* most students know there are 100 cm in a meter, but have difficulty in answering the question "How many cm^2 are in a m^2?" or "How many cm^3 are in a m^3?" Converting Fahrenheit to Celsius causes difficulty for many students. Also, many students have difficulty estimating distances, temperature, and mass.

G. Factors contributing to students' difficulties, confusion, and misconceptions

Many students, even at the college level, are concrete thinkers. Converting units and making sense of the units for area and volume require an understanding of ratio and proportion, which many students do not possess. Some measurement activities require an understanding of fractions, decimals, and algebra, which also cause difficulties for some students.

Textbook and classroom presentations of concepts related to measurement do not usually take into account student readiness and make assumptions about students' entry level that do not match many students.

H. Appropriate teaching strategies

The measurement activities lend themselves well to conceptual change, discussion and demonstration, and collaborative learning strategies. As a teacher, you should use your own judgment as to the appropriateness of a particular strategy for each activity, although we have some suggestions. We suggest that collaborative learning be used with activities 4, 5, 6, 8, 9, 10, 12, 13, and 14. Activities 1, 2, 3, and 11 may be approached with the conceptual change strategy. Discussion and demonstration may be used with activities 1, 2, 5, 7, 14, and 15.

I. Teaching notes

The measurement activities are designed to meet the needs of students of various ages. Some of the activities are intended primarily for elementary-level students, while others are more appropriate for secondary students. Still other activities may be used with K-12 students. Activities 1, 2, 3a, 3b, 3c, 7a, 7b, 8, 11, and 16 may be used with elementary school children. Others are suggested for secondary students.

J. Materials

    rope licorice
    jar of M & M candies
    grid or graph paper
    scissors
    masking tape or cellophane tape
    containers of water of different temperatures
    washers
    thermometers
    balances
    measuring tapes
    rulers with mm and cm markings
    calipers
    micrometers
    calculators

K. Activities

See the following pages.

1. Need for a standard: " Message from the aliens"

NEWS ALERT!

"TOMORROW EVENING AT 8:00 EST
THERE WILL BE AN IMPORTANT MESSAGE.
STAY TUNED TO THE NATIONAL CHANNEL."

WAIT
wait
wait

At 8:00 EST the message arrives:

"This is the voice of the national radio. We have with us a university professor who is expert in decoding messages. We have been told that we should expect, at any moment, a message from another planet, and ...
Here, I think we are receiving the message! Professor, please help us make sense of the message."

The professor begins working to interpret the sounds which are being received. Finally, the professor has been able to decode the message and is ready to share it with us.

"This message comes to you from planet @*. We have in our possession one of your baseball stars. We have conducted a variety of tests on him and know a great deal about him. If you want to have him back, we are demanding that you send to us plans and the dimensions of your school, since we want to build schools similar to yours.

However, you need to know that we know nothing about your measuring system and units. You have a little time to get back to us and, again, use the decoder to give the message to us."

WHAT SHOULD WE DO? HOW SHOULD WE PROCEED?

Get into your small group and come up with a plan.

Share your plans with others.

Which plans are feasible to accomplish this mission? Why?

Implement the plan. Using the units selected, measure several items in the room, such as a desk, blackboard, filing cabinet, and the room itself.

Report your results to the rest of the class.

When the class is done with the project, the results are given to the professor, who encodes the message and sends it. Then the class waits on the fate of baseball player.

WAIT
wait
wait

Finally, a message arrives from @*, which the professor decodes:

"We don't have confidence in your figures. We are very confused. We also have had in our possession a famous Japanese baseball player, and the figures we received from his country were totally different from yours."

WHAT IS THE PROBLEM? HOW CAN WE DEAL WITH THE PROBLEM? WHAT HAVE WE DONE HERE ON EARTH TO DEAL WITH THIS SORT OF PROBLEM?

In your group, get a meter stick and measure the same items. Report the results to the class. What is the difference between this set of measurements and the first set?

What do we learn from this activity about the importance of having a standard for measurement?

With what standards of measurements are you familiar?

2. Non-standard measurement

Use your fingers to measure a piece of licorice. How many fingers long is it? Also use your partner's finger to measure it.

If the teacher were to give you five finger lengths of licorice, whose finger would you want used to measure it--yours or your partner's? Why?

If you could choose anyone in the room to measure out six fingers of licorice, who would you choose? Why?

How could you make sure that everyone in the room got the same amount of licorice?

This same activity could be done with measurement involving mass and time.

3. Estimating

Given each of the following situations, estimate and give reasons for your estimate, then suggest ways to check your estimate.

a.   Number of M&M's in a jar

b.   Length of the room

c.   Area of a table

d.   Volume of a box

e.   Temperature of tap water

f.   Mass of a 707 jet

Share your estimates and plans on ways to test your estimates with others in the class.

4. How much material is needed to COVER something? How much stuff do we need to FILL something?

Given a book, a desk, and the blackboard, work in small groups to design a way to decide how much material would be needed to cover each thing.

Given boxes of various sizes, determine some effective and quick ways to decide what is needed to cover each box. What statement can you make about information that is necessary to determine how much material is needed? What units do we use?

Get sheets of graph paper, scissors, and tape. Build boxes of various sizes. Decide on ways you can determine how much stuff would be needed to fill each box. Can you come up with an efficient way to determine the volume of a box if we know the dimensions? What units will you use?

5. Conversions

Get appropriate measuring tools and determine the following the conversions:

a.   how many cm in 2 m

b.   how many mm in 20 cm

c.   how many mm in 2 in

d.   how many cm^2 in 3 m^2

e.   how many gm in 4 KG

f.   how many in^2 in 1 m^2

g.   how many mm^3 in 2 dm^3

h.   how many degrees F = 20 degrees C

i.   how many degrees C = 80 degrees F

6. How much do we rely on our senses?

a.   Temperature

Get a container of warm water from the tap. Estimate the temperature and record your estimate, then measure the temperature with a thermometer. How did your estimate compare with the measurement?

Get a container of cold water and a container of warm water. Have someone in the group measure the temperature of the water in each container and not reveal their measurements to you. Then place your hand in the cold water and leave it there for a few seconds. Remove your hand from the cold water and place it immediately in warm water. What do you estimate the temperature of the warm water to be? Record your estimate, then check with your friend who had the measurement. How did your estimate compare with the measurement?

This time place your hand in warm water and leave it for a few seconds. Remove your hand and place it in the cold water. What is your estimate of the temperature of the cold water? Record and check your estimate with the actual measurement.

What statements can you make about this experiment?

b.   Reflexes: How quick are we?

If a partner asked you to predict your reaction time, what would you respond?

Check your reaction time by trying to catch a dollar bill or similar paper between your thumb and index finger as your partner releases it.

7. Measurement at home

Keep track of the measurements which occur in your home during a 24-hr period.

Bring the information you have collected to class to share with others.

8. Using appropriate tools and units

You are asked to make accurate measurements of each of the following. In your small group design a plan on how you will go about making the measurement and what appropriate tools you will use. Share the process and the rationale you used to select the measuring tool and units.

a.   mass of a washer

b.   temperature of the room

c.   mass of a sheet of paper

d.   mass of the water in a full bath tub

e.   volume of the room

9. Indirect measurements

In your group, design a way to measure the height of the school flag pole.

Share your plan with others in the class.

Get the necessary material and determine the height of the flag pole.

Compare your numbers with those of others.

If you were to do this over, what would you do differently?

What are some examples of indirect measurement with which you are familiar?

10. Creating your own standards

You are familiar with standards people use to measure length, mass, time, speed, temperature, energy, power, etc. The challenge for your group is to create your own standard of measurement for some fundamental units.

Then devise derived units using your standards.

Share your units with others.

11. Describing objects accurately

You have access to various objects. Without naming the object or mentioning its color or other obvious attributes, use measurements such as volume, density, conductivity, and others to describe the object to other groups in the class. See if they can follow your measurements and choose the item you have described.

12. Accuracy and precision

Use a vernier caliper, a ruler that reads to mm, and a ruler that reads to cm.

Working in groups of 3, have each member of the group measure the diameter of a rod, using each of the three instruments provided.

Each member would measure the rod with the vernier caliper to the nearest 1/100 mm.

Each member would measure the rod with the mm ruler to the nearest 1/10 mm.

Each member would measure the rod with the cm ruler to the nearest mm.

Calculate the variation in the measurements made by your group. To do this, write the largest and smallest readings taken with each instrument and find the difference between these readings. These are maximum variations for the measurements made with each instrument. Compare the variations of your group with those obtained by other groups.

What caused this variation?

What is your error of measurement? The error of measurement for each instrument is ± one half the smallest scale division that can be read with an instrument.

How should we determine error of measurement for the entire class?

13. Tolerance in measurement

You will need a measuring tape, a ruler with both mm and cm markings, and a calculator.

Have each group measure the dimensions of the same desk to the nearest 1/16 in.

Draw a rectangle having dimensions of 25 cm by 15 cm. Measure the dimensions of this rectangle with the tape measure to the nearest 1/16 <u>in</u>.

Determine the tolerance range of your measurements.

Calculate the smallest possible area of each rectangle.

Calculate the largest possible area of each rectangle.

What is the upper limit and lower limit of each area? How do the values compare for the two rectangles?

If you measured the length and the width of the classroom floor to the nearest 1/16", how do you think the range for the area between the upper and lower limits for measurements of the floor would compare to the ranges for the desk and the drawn rectangle?

14. Measuring things that appear to be unmeasurable

In your group design ways to make the following measurements:

a.   distance around a pond
b.   mass and volume of a bean
c.   mass and thickness of a sheet of paper
d.   volume of a pumpkin

Implement the plan by making the measurements.

Report your results to others.

Find examples of professionals who use similar strategies to make measurements of things that are not directly measurable.

15. Your own unmeasurables

Bring something that cannot be measured and design a way to measure it.

L.   Assessment ideas

Pencil-and-paper tests can be used to test students' knowledge of fundamental information about measurement units and conversion.

Students may be assessed on their projects and investigations. These assessments may be useful in allowing students to communicate their questions and problems, their plans for their project, the implementation of their plans, and their results.

Observation of students performance, individually or in groups, enables assessment of their skill to use instruments, collaboration with others, ability to estimate and test their estimation, and ability to make sense of their data.

Interviewing students individually or in small groups can be particularly useful in gaining insight into student thinking, the difficulty she/he may be encountering, and the changes taking place in your students' knowledge and skills.

Portfolios can be powerful assessment tools that are useful for students, teachers, and parents.

M.   Resources and references

Center for Occupational Research and Development. (1988). *Measuring in English and Metric Units.* Waco, TX: Author.

Center for Occupational Research and Development. (1988). *Precision, Accuracy and Tolerance..* Waco, TX: Author.

National Council of Teachers of Mathematics. (1989). *Curriculum and Evaluation Standards for School Mathematics.* Reston VA: Author.

10
RATIO AND PROPORTION

A. Overview

This chapter includes definitions, uses, and importance of ratio, proportion, and scale.

B. Background information for the teacher

Ratio, proportion, and scale

A *ratio* is a fraction that expresses a relationship between two numbers. The numerator denotes how many pieces of the whole or parts of the set are present and the denominator represents how many pieces or parts make up the whole. A *proportion* is a statement that equates two ratios. For example, 3/4 is equivalent to hitting 75 out of 100 pitches, or 75%.

A *scale* is a ratio that expresses a measurement relationship between two numbers. Maps are usually drawn to a specified scale, as when 1 inch is equal to 15 miles (1:15); thus, a path on the map that is 3 inches long represents 45 miles. Other scaling examples are enlarging a photograph or creating a scale model of a building, car, or train.

Who uses ratio, proportion, and scale?

Many professionals use ratio, proportion, and scale in their everyday work, including photographers, cartographers, artists, carpenters, builders, architects, interior decorators, designers, and manufacturers. In daily life, ratio and proportion are used to scale recipes down or up, figure weather probability and sports statistics, read maps, and understand data referred to in news reports.

Why is it important for students to learn ratio and proportion?

Failure to develop proportional reasoning by the end of the middle grades inhibits students in many other areas. Algebra, geometry, calculus, statistics, biology, physics, chemistry, medicine, pharmacy, sociology, marketing, resource management, environmental regulation and planning, economics, and many other disciplines require this type of quantitative reasoning. Without proportional reasoning, students cannot adequately analyze problems that involve comparisons and covariance, let alone such citizen issues as those involving the ideas of "fair share," democratic representation, pollution, and taxation.

C. NCTM position on ratio and proportion

K-4

Children will begin to represent and describe mathematical relationships.

Patterning is a logical precursor to understanding ratio and proportion.

A concrete understanding of fractions and decimals is also important.

5-8

Students will understand and apply ratios, proportions, and percents in a wide variety of situations.

Students will begin to reason proportionally.

9-12

The study of proportional reasoning will receive increased attention, as it is critical both as a foundation for further mathematics study and the application of mathematics.

D. Integrating problem solving, reasoning, communicating, and making connections

Ratio and proportion are excellent sources for problems that involve real-life situations and reasoning; for instance, rate problems and currency problems incorporate ratio, proportion problems, and reasoning. These sorts of problems have many connections to other disciplines as well as other parts of mathematics, described earlier. Communication of ratio and proportion situations is frequently used in everyday life. Students can analyze newspaper examples and bring in examples from their own experiences and interests.

E.   Prerequisite skills and knowledge

Students must understand the operations of multiplication and division, as well as concepts related to fractions, to make sense of ratio and proportion.

F.   Students' difficulties, confusion, and misconceptions

Students often interpret enlargement as the addition of a fixed amount rather than thinking of it as multiplication by a scale factor (the addition strategy error). For example, when given the first shape below and asked to create the remaining side of the next shape to make it the same shape but bigger, students who have the addition error strategy would say that the missing side should measure 3 because $3 + 2 = 5$ and $1 + 2 = 3$.

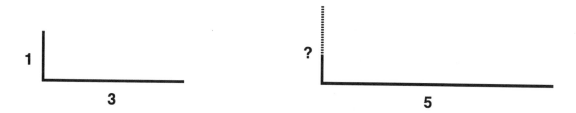

G.   Factors contributing to students' difficulties, confusion, and misconceptions

While a variety of researchers have identified the addition strategy error (Hart, 1984; Karplus, Karplus, Formisano and Paulsen, 1975; Piaget and Inhelder, 1967), the cause seems to be interpreted by some as a developmental progression that most learners pass through (Piaget and Inhelder, 1967). Others interpret the error as something that can be approached without the addition error (Karplus, et.al., 1972) if students have an adequate grasp of the concepts of fractions and the multiplication of fractions (Hart, 1984).

H.   Appropriate teaching strategies

Appropriate teaching strategies for these topics include discussion and demonstration, use of discrepant events, mental model building, and teaching for conceptual change. In activities such as "Traveling in Europe" discussion and demonstration are important so students can demonstrate their understanding of conversions. Converting money and measurement is an important part of ratio and proportion that students need to discuss, not just work exercises. Pay special attention to asking "Does this make sense?" so students can evaluate their ideas.

Finding movie inconsistencies in the "Honey, help me get it right" activity uses the discrepant event strategy. Mental model building is appropriate for "Scale it!", in which students should be able to picture dilations of a figure using a given ratio. "Exploring levers" uses the teaching for conceptual change strategy, in which students are encouraged to examine and test their original ideas (preconceptions).

I. Teaching notes

Black and white outline drawings work best for "Scale it!" because they are easier to replicate and they eliminate the distracting variables of color and shading from the scaling lesson.

Some suggested movies that will work well for "Honey, help me get it right!" are: *Honey, I Shrunk the Kids!*, *Honey, I Blew Up the Kid!*, and *Honey, I Shrunk the Kids Again!*. All are Disney films and are rated PG. A middle school audience would probably be the most motivated to watch the movie. If you think a class might be inattentive to watching one of the movies for the mathematics content, you might require them to record a certain number of proportion observations. Older students might enjoy gathering observations from science fiction movies in which scale models of creatures and spacecraft are used.

"Traveling in Europe" requires the knowledge that 1 liter is equivalent to 0.2642 gallon or 1 gallon = 3.785 liters. Students can also look up currency conversion tables in the financial section of the newspaper or on the Internet to create more up-to-date situations.

J. Materials

      appropriate movie, as mentioned in Teacher Notes above
      rulers and/or yardsticks/meter sticks
      tape measures
      large grid paper
      small grid paper
      interesting black-and-white drawings; e.g., clip art, cartoons, other illustrations
      newspaper with currency conversions
      scales and weights
      computer with a spreadsheet program

K. Activities

See the following pages.

1. Scale it!

You can create scale replicas by using a scale factor. For example, the picture on the left can be enlarged or shrunk by scaling it up or down. To do this, divide all of the picture into squares, as shown.

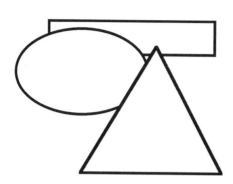
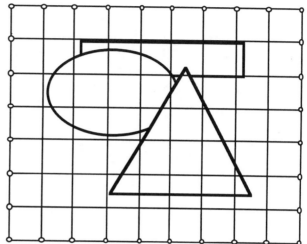

Then enlarge or shrink the picture one square at a time, creating the same part of the picture for each square on a smaller or larger square grid.

Demonstrate that you can do this by creating this picture in a 4″ x 4″ space.

Is it possible to enlarge this picture to any other dimensions? Are there any criteria for the dimensions that might be used? Explain.

How do the dimensions affect the picture?

Choose a black line comic (one frame) and enlarge it using the same procedure.

Explain why this enlargement technique works.

What are some applications of this technique that could be used in conjunction with school or extracurricular activities?

2. Honey, help me get it right!

Watch the film and write down any actual references or clues you notice related to measurement, scale, and proportion while you are watching the film. Be sure to note any scale that appears to be inaccurate.

a.   In *Honey, I Blew Up the Kid!* the hands of the toddler hold the teenagers when the toddler is 50 feet tall. Is this reasonable? Let's assume the teenagers are 5 feet tall. Measure your palm from pinky to pointer fingers (which is how they were held in the movie). What do you think? Is the scale shown in the movie scene reasonable or not? Explain.

b.   In another scene of *Honey, I Blew Up the Kid!* the police officer claims that a fire truck ladder wouldn't reach beyond the toddler's knees. How high would you expect the ladder for the biggest fire truck in Las Vegas to extend? Is it reasonable for it to only reach to the toddler's knees when he is 50 feet tall? 100 feet tall? Explain your reasoning.

c.   Using your observations, create a proportional-reasoning problem for yourself using the film you watched. Solve this problem and explain your reasoning. Was the film valid in this situation?

Share your problem with another team and solve their problem while they try to solve your problem.

3. Traveling in Europe

Suppose you are traveling in Europe and need to convert money from one currency system to another as you go from one country to the next.

If the exchange rates are:

$1.60 U.S. = 1 British pound
$0.25 U.S. = 1 French franc
$0.71 U.S.= 1 German mark
$1.00 U.S. = 1580 Italian lire

a.   Which would you rather have, 1000 pounds or 1000 lire? Why?

b.   If you have 25 francs, what do you have in dollars?

c.   If you find gas for 4 francs per liter, how much would you be paying in dollars? per gallon? Is this a good price? Explain.

d.   How many dollars is 17 marks worth?

e.   You travel from England to France and want to exchange 12 pounds. How many francs will you receive?

f.   You paid 1500 lire per liter for gasoline. Is this a good deal? Explain your answer in terms of dollars per gallon.

g.   When you get back from your European vacation, you have 2330 lire, 5 pounds, 8 marks, 35 francs, and 35 dollars left over. How much is this worth in dollars? In francs? In pounds?

h.   Create a conversion problem and solve it.

i.   Have your neighbor solve your problem while you solve your neighbor's problem. Show how you solved it.

j.   Try to create a spreadsheet to help answer any of the conversion problems for the currencies given.

4. Mixing Gorp

The cookbook that you are using has ingredients listed by ratio to make it easy to adjust the size of the batch you'd like to make.

Here is the recipe for Gorp you have chosen to use:
3 parts peanuts
1 part chocolate chips
2 parts dried apple bits
1/2 part cashews

Sometimes you may have a certain amount of one ingredient available, and want to adjust the recipe to fit that ingredient. For each of the situations below, complete the appropriate measures for the remaining ingredients in each batch.

a.   You have 1 cup of chocolate chips.

b.   You have 1 cup of dried apple bits.

c.   You have 1 cup of cashews.

d.   You have 1 cup of peanuts.

e.   You have only 1/2 cup of peanuts.

f.   You have 3/2 cup of peanuts.

g.   You have 2/3 cup of chocolate chips.

h.   You have 3/4 cup of chocolate chips.

i.   You have 1/2 cup dried apple bits.

j.   You have 3/4 cup of cashews.

5. Exploring levers[4]

Look at the measuring balance with weights on each side of the fulcrum represented in the figure below.

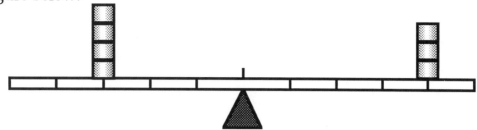

Predict which of the following will happen if you set up the situation above:

    a. The weights will balance.

    b. The lever will tip down to the right.

    c. The lever will tip down to the left.

Give an explanation for your prediction.

Share your prediction and explanation with the others in your group.

Have a representative share all of these ideas with the class.

In your group, get the necessary materials and set up the situation in the figure. What happened?

Did your observations agree with your predictions?

Why do you think they did or did not?

Do you want to make any changes in your thinking about levers at this point?

What does the relationship of the weights and the distance from the center have to do with the balance?

Test out your ideas with the materials provided.

Make a conjecture about these relationships that you can prove to the class. Write your conjecture.

Relate your observations and conclusions to the concepts of ratio and proportion.

---

[4] This activity is adapted from Stepans, J. (1994). *Targeting Students' Science Misconceptions: Physical Science Activities Using the Conceptual Change Model.* Riverview, FL: Idea Factory.

L.   Assessment ideas

Thought-provoking problems require students to think about how the concepts are working.

Journaling encourages students to record not only how they feel the mathematics is working, but also provides information to document how their ideas change over time.

Creating a scale model (*e.g.*, the classroom, the school, their neighborhood, a map of the town, or an object) allows students to demonstrate their understanding of both ratio and proportion.

M.   Resources and references

Hart, K. M. (1984). *Ratio: Children's Strategies and Errors*. Windsor, Berkshire, GB: Nfer-Nelson.

Karplus, R., Karplus, E., Formisano, M., and Paulsen, A. C. (1975). *Proportional Reasoning and the Control of Variables in Seven Countries*. Advancing education through science programs, Report 1D-65. Berkeley, CA: Lawrence Hall of Science.

National Council of Teachers of Mathematics. (1989). *Curriculum and Evaluation Standards for School Mathematics*. Reston VA: Author.

Piaget, J. and Inhelder, B. (1967). *The Child's Conception of Space*. London: Routledge & Kegan Paul.

Stepans, J. (1994). *Targeting Students' Science Misconceptions: Physical Science Activities Using the Conceptual Change Model*. Riverview, FL: Idea Factory.

11
DISCRETE MATHEMATICS

A. Overview

Concepts included in this chapter are: discrete vs. continuous mathematics, existence problems, counting problems, optimization problems, applications, permutations, combinations, networks, and Euler and Hamiltonian circuits.

B. Background for the teacher

What is discrete mathematics?

The NCTM Standards defines *discrete mathematics* as the study of mathematical properties of sets and systems that have only a finite number of objects. For example, discrete mathematics problems would include such real-life problems as organizing a schedule, figuring the number of possibilities for a telephone number, finding limitations of a four-letter password coding scheme, determining the best possible place for a road, and constructing a viable network for shipping packages.

To understand the difference between discrete and continuous mathematics concepts, compare two clocks. An analog clock has a face with hands that go around so time can be measured continuously. A digital clock tells time to the nearest minute or second, so you don't know how far into the minute or second you are. It has discrete times. Thermometers also are either analog or digital. A digital ear thermometer reads as a decimal to the nearest tenth of a degree. A traditional oral thermometer gives a continuous reading between markings.

What types of problems constitute discrete mathematics?

Dossey (1991) separated discrete mathematics into three categories: *existence* problems, *counting* problems, and *optimization* problems. A maze problem can demonstrate these three categories, as shown in the following illustrations.

Here is the maze we will use to demonstrate the three types of problems in discrete mathematics:

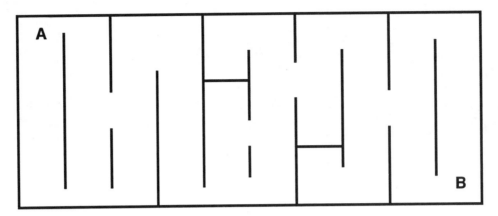

Question 1: Can you get from A to B without crossing any lines?

Existence problems deal with whether a problem has a solution or not. If you can prove a solution exists, then the problem is solved. So, if you are able to get from A to B then you have demonstrated that a solution exists, as shown below.

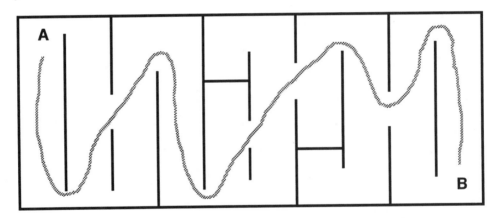

Question 2: How many different paths can you take to get from A to B without crossing back over your path?

Counting problems investigate how many solutions are possible to a problem where it is known that a solution exists. Refer to the next maze diagram. At point A you can choose from two viable paths; at the ● you can go either of two ways; the ▲ also offers two choices, as does the ■. All of these choices can affect the others, creating different path choices at each fork and therefore a variety of trips.

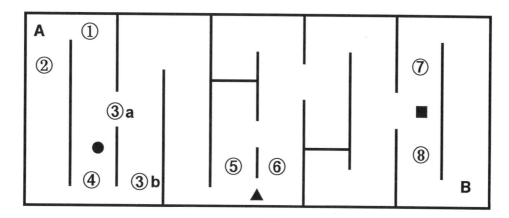

Using a tree diagram reveals 16 possible paths through the maze.

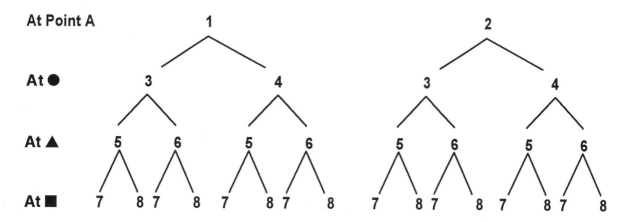

Question 3: What is the shortest path?

The *optimization problem* asks for the best solution to a particular problem. So in this maze problem, an optimal solution would be the 1-3-5-8 or the 1-3-6-8 path. Both are shown below. The shorter distance for the maze is to choose 1 to 3 and then, at the ■, choose 8. Optimization problems are often the most interesting.

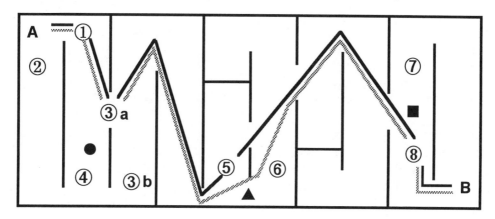

How is discrete mathematics used?

The use of discrete mathematics has grown steadily since the 1980's due to the many business and industrial applications of computer science, which depends on discrete mathematics. As computers have become more powerful, it has become possible to explore larger discrete mathematics problems.

Anyone who knows the possible outcomes or choices that need to be calculated can use discrete mathematics. Computer programmers, managers, marketers, stock brokers, telephone companies, shipping companies, and anyone who is trying to set up a network all use discrete mathematics. Specific examples of use include organizing an employee work schedule, figuring deadlines for completing homework and projects, finding optimal phone line placements, calculating the possibilities for an ATM password, deciding the format of your state license plates, and encoding messages on the Internet to protect privacy.

Permutations and combinations

In *permutations* all items are selected from the same set, no item may be used more than once, and the order is important. To solve a simple permutation problem you just need to think about the number of items to be ordered (n) and the possible arrangements are n factorial (n!).

Imagine that you are putting together a golf foursome and want to know if you can change the order of tee off at each hole and not repeat the same order through eighteen holes of golf. To figure this out, you assume that any of the four golfers can be the first to tee off. Then second position can be filled by any of the three remaining golfers and either of the remaining two golfers could be the third start. The fourth start has only one remaining player possibility. The total number of possible arrangements is 4 x 3 x 2 x 1 = 24 because there are four to choose from in the first position, three the next, and so on. So, in answer to the question, it would not be necessary to repeat playing order in 18 holes. The solution could also be expressed as 4! = 24 using factorial notation:

$$n! = n \bullet (n-1) \bullet (n-2) \bullet (n-3) \bullet ... \bullet 3 \bullet 2 \bullet 1$$

Combinations are selections of items made when the order is not important. Imagine I want three scoops of ice cream and there are five flavors. It doesn't matter what order the scoops are put into the dish, so calculating the possibilities will be a combination problem. To answer the question, I would take the five choices, such as Vanilla (V), Chocolate (C), Strawberry, (S), Lemon (L), and Bubble gum (B), and arrange them in all possible three-choice solutions. Since VCS is equivalent to CVS, there is no need to count them twice. This means we have: VCS, VCL, VCB, VSL, and VLB where vanilla is one of the scoops. When chocolate is one of the scoops and vanilla combinations are not included, the possibilities are CSL, CSB, and CLB. The last possibility without repeating any of the chocolate or vanilla scoop possibilities is SLB. Therefore, there are 6 + 3 + 1= 10 possible choices. For many problems there are too many possibilities to calculate in this way, so a formula is used for calculating the combinations for r items selected from n items:

$$_nC_r = \frac{n!}{(n-r)! \cdot r!}$$

or for our problem:

$$_5C_3 = \frac{5!}{(5-3)! \cdot 3!} = \frac{5 \cdot 4 \cdot 3 \cdot 2 \cdot 1}{(2 \cdot 1)(3 \cdot 2 \cdot 1)} = \frac{20}{2} = 10$$

What do combinations and permutations have to do with discrete mathematics?

When the possible answer is a discrete solution, the problem could be included in a discrete mathematics unit. Since most such problems call for a finite answer, they are excellent sources for discrete mathematics problems. Permutations are used when the order is important and combinations are used when the order is irrelevant.

Networking problems, Euler circuits, and Hamiltonian circuits

Networks are used to model connections between discrete points. Transportation problems are a good example of networking problems. For instance, if a mail carrier wanted to travel most efficiently and be able to deliver all of the mail on the route, that would be a networking problem. Telephones, computers, planes, trains, rivers, cities, etc., are all rich with networking problems.

Mathematics networking problems include Euler (pronounced "oiler") and Hamiltonian circuits. An Euler circuit is one in which each connecting line segment is traversed only once. In a Hamiltonian circuit each vertex is visited only once. In the network shown below, an Euler circuit would be **a-b-c-g-f-d-e** and a Hamiltonian circuit would be **A-B-C-D-E**. Note that both types of circuits require travel on existing paths, but in a Hamiltonian circuit it is not necessary to travel <u>every</u> path. Both of these types of circuits are useful for solving problems involving graph theory.

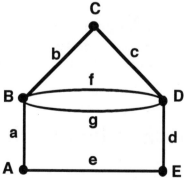

C. NCTM position

K-4

Patterning is a logical part of doing discrete mathematics and should be taught so students might recognize patterns in discrete mathematics.

The existence of answers should be demonstrated.

5-8

The array and combination representations of multiplication should be taught so students can understand the multiplication principle, a basis for concepts in probability.

9-12

Discrete mathematics are specifically addressed.

Students will represent finite graphs, matrices, sequences, and recurrence relations.

Students will develop and analyze algorithms.

Students will be able to solve enumeration and finite probability problems.

D. Integrating problem solving, reasoning, communicating, and making connections

Questions in discrete mathematics usually call for problem solving, involving intense thinking. Reasoning is also necessary for doing discrete mathematics if teachers do not simply give formulas for solving the problems. Students can usually reason out solutions in discrete mathematics problems. Since it is important to explain the reasoning used to solve these types of problems, communicating is also a critical component. The connections to a variety of real-world situations are numerous. It is a relatively new field whose importance is being underscored by the emerging technologies in computers, communication, and transportation.

E. Prerequisite skills and knowledge

Most of the visual discrete problems do not involve any kind of operation experiences, just analytic ability. Students can solve these sorts of problems at an early age and the learning can be reinforced by adequate questioning. To learn permutations and combinations, students should have an understanding of multiplication.

F. Students' difficulties, confusion, and misconceptions

Ironically, a possible difficulty might be the misperception that discrete mathematics isn't *real* mathematics because it is fun. Children do maze problems, color, and play with sequencing for fun at home. Formalization of these sorts of problems and the analysis of the problems are important. It might be difficult for children (and some parents) to realize that people do this as their work.

Students also may confuse permutations (order-dependent) and combinations (order-independent).

G. Factors contributing to students' difficulties, confusion, and misconceptions

Discrete mathematics is a topic that many teachers use only as an enrichment. It is not frequently taught as part of the curriculum.

Permutations and combinations are often taught at the same time and students memorize the formulas without developing the understanding of the notation or the concepts involved.

H. Appropriate teaching strategies

Mental model building, discussion and demonstration, and teaching for conceptual change are effective strategies for teaching discrete mathematics. Mental models are an important part of discrete mathematics. The questions stressed in Teaching Notes should help students create models for understanding discrete mathematics problems. "The garden" activity, for example, is made easier by creating a mental model. Discussion and demonstration is an excellent strategy for "A maze-ing," "Order in the court," and "Connections?" activities, encouraging students to share ideas on simplifying the problems. The conceptual change strategy used with "How many colors?" causes students to predict solutions and explore the problem in depth.

I.   Teaching notes

Gardiner (1991) cautions
> "The educational value of much simple discrete mathematics lies precisely in the fact that it forces students to *think* about very elementary things, such as systematic counting. However, this can easily be undermined by the fact that most mathematics teachers feel obligated to 'help' students solve hard problems by reducing the number of manageable and predictable steps, or rules, *requiring an absolute minimum of thinking*."

<u>It is important for teachers to counter this natural "help" tendency and let students THINK THROUGH the problems.</u>

There are many excellent sources for maze problems. Begin with easy maze problems, and pose such "thinking" questions as:

> Is it possible?
> How many different solutions are possible?
> What is the shortest solution?
> What is the longest solution?

The level of difficulty can be increased with more complex maze problems, but these questions are important ones for all students to think about. We hope students will begin to automatically ask such questions so they can develop a schema for maze problems. Students should also begin developing their own maze problems early.

When dealing with Euler and Hamiltonian circuits, note the important generalization of Euler: *All of the vertices must have an even number of paths for an Euler circuit to be possible.* So, when solving a network problem, visualize the network to see if a solution is possible. For example, the garden path problem in the activities can be reduced to the following network:

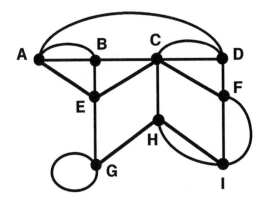

This network schematic helps us to see that all of the vertices have an even number of paths from which to choose. It also makes the problem less complex.

J.   Materials needed

graph or grid paper
crayons or colored pencils
maps (preferably uncolored)

K. Activities

See the following pages.

1. A maze-ing activity

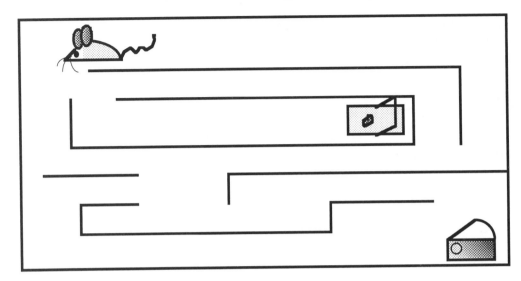

Demonstrate how it is possible for the mouse to get to the cheese.

How many different ways is it possible?

What is the shortest path?

Create a maze through which there is only one way to get from point A to point B.

Try creating a maze that can have more than one solution.

Try creating a maze that has more than one solution __and__ has a shortest route.

Share your maze with a partner and see if she or he can solve it.

2. How many colors?

Use a pencil to draw a figure that begins and ends at the same point. You may not lift the pencil while drawing the figure.

What is the least number of different colors it will take to color in the figure you created so that all of the spaces in the figure can be recognized as distinct?

Can you create a figure, following the rules given, that will take more than three colors to color it in? Demonstrate and explain why you can or cannot do it.

Make a conjecture about the minimum number of colors you think it will take to color any figure, such as a map.

Share your ideas in your group and discuss the possible validity of everyone's conjectures. Modify your idea as needed based on the discussion.

Convince me to acknowledge or reject the validity of your conjecture.

Now overlap two closed figures by drawing one on top of the other. What is the fewest number of colors it will take to color this figure to show all of the distinct areas created?

Make a conjecture about the minimum number of colors you think it will take to color any two overlapping closed figures.

Share your ideas in your group and prove or disprove your conjecture.

What do you predict will happen if you try overlapping three closed figures?

3. Connections?

Imagine that your county is setting up a new cable system to connect several towns to a new cable television hookup. The arrangement of the six towns (A-F) is shown, along with the mileages between towns. (Note that the picture is not drawn to scale.) The headquarters with the satellite dish can be located in any one of these cities.

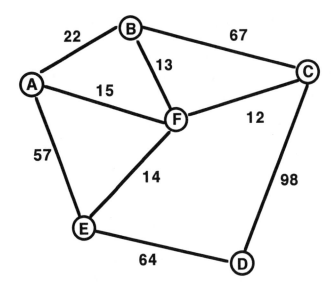

Is it possible to connect all of the cities to cable?

What would be the minimum amount of cable needed if you can only bury it on the easements (paths) provided?

Explain your reasoning.

Would it be best to locate the headquarters at one particular city over the others?

Why or why not?

If it costs three thousand dollars per mile to string the cable, what would your savings be over the most expensive route?

Explain your reasoning.

4. The garden

You decide to tour the garden, whose paths are shown in the diagram below, but you know yourself well enough to know that you are easily bored. Plan a walk so that you do not go over the same garden path twice (except at intersections) but also get to travel every path. Such a walk is called an Euler circuit.

Imagine you collect $50 for each intersection in the garden. Unfortunately, you only get the money the first time at the intersection and when you pass over an intersection you've already been on, it is impossible to get any more money. What route would you take to earn the most money with the least walking?

Is it possible to begin anywhere in the garden and end at the same spot without walking more than once over the same path, but traveling on every pathway?

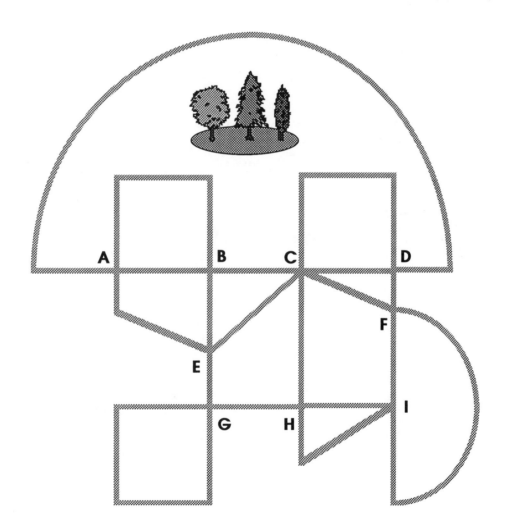

5. Order in the Court

A very democratic jury decides to walk into the courtroom a different way every day. They choose to do this by changing the order they single-file into the courtroom daily.

If there are six jurors on the jury, how many days can they walk into the courtroom in a different order?

Explain your reasoning.

There are 10 prospective jurors in the waiting room. How many different six-person juries could be formed?

Explain your reasoning.

There are 4 witnesses that the attorney wants to call to the stand. In how many different possible orders can she call them to the stand?

Explain your reasoning.

Imagine an attorney had six witnesses. Since they all had the same thing to say, he only needed four to prove his case.

How many different ways could he call his four witnesses?

Explain your reasoning.

L. Assessment ideas

Performance assessment strategies that would be appropriate for discrete mathematics include maze problems with questions such as those suggested in the teacher ideas, and the four-color map problem.

A portfolio collection depicting successful strategies for solving discrete mathematics problems and journal writing about problems that explain the strategies used to solve discrete mathematics problems allow students to express their reasoning and depth of understanding.

M. Resources and references

Bennett, J. O., Briggs, W. L., and Morrow, C. A. (1996). *Quantitative Reasoning: Mathematics for Citizens in the 21st Century.* Reading, MA: Addison-Wesley.

Dossey, J. A. (1991). Discrete mathematics: The math for our time. In M. J. Kenney & C. R. Hirsch (Eds.) *Discrete Mathematics across the curriculum, K-12 , 1991 Yearbook.* Reston VA: NCTM.

Gardiner, A. D. (1991). A cautionary note. In M. J. Kenney & C. R. Hirsch (Eds.) *Discrete Mathematics across the curriculum, K-12 : 1991 Yearbook.* Reston VA: NCTM.

National Council of Teachers of Mathematics. (1989). *Curriculum and Evaluation Standards for School Mathematics.* Reston VA: Author.

12
SPATIAL THINKING

A. Overview

Concepts included in this chapter are: spatial thinking, eye-motor coordination, figure-ground perception, perceptual constancy, position-in-space perception, perception of spatial relationships, visual discrimination, visual memory, transformational geometry, mappings, symmetry, slides, and rotations.

B. Background for the teacher

What is spatial thinking?

Spatial thinking is reasoning using spatial relationships that have been recognized and interpreted. It involves the transformation and manipulation of constructed spatial objects, which can be in the form of mental or physical representations. Del Grande (1987) named seven abilities that are important in spatial perception: eye-motor coordination, figure-ground perception, perceptual constancy, position-in-space perception, perception of spatial relationships, visual discrimination, and visual memory.

Eye-motor coordination and figure-ground perception

Coordinating vision with the movement of the body uses the *eye-motor coordination* ability, as when you trace over a dotted circle. Children who have difficulty with a simple motor skill will have trouble concentrating on anything besides the motor skill. Only when the motor skill is habit will the child be able to concentrate on other parts of the learning.

Figure-ground perception is being able to ignore extraneous markings in a picture to perform a task. For example, in the following diagram, "Trace around the rectangle." and "Color the circle purple." would be tests to see if students are able to ignore the extraneous shapes. It is important for children to be able to break down figures and reassemble them to learn to find embedded figures.

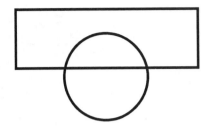

Perceptual constancy and position-in-space perception

Objects have invariant sizes and shapes. For example, a cube can be recognized from a various viewpoints. The ability to recognize an object and realize that it does not change size or shape when viewed from a variety of views is termed *perceptual constancy*.

The ability to determine the relationship of objects is termed *position-in-space perception*, and it is important in recognizing patterns. For example, positioning a red cube between two blue cubes at an elevated position and explaining the relationship among these three cubes would be demonstrating position-in-space abilities.

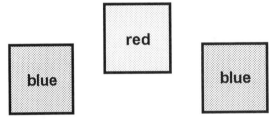

Perception of spatial relationships, visual discrimination, and visual memory

When students can explain the relationship of two or more objects in relation to each other, they are exhibiting a *perception of spatial relationships*. This perception is often closely related to position-in-space perception. For instance, a student demonstrates a perceptual of spatial relationships when noticing a congruence transformation for two objects involving a flip, slide (shown below), or rotation.

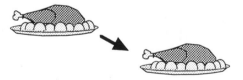

Visual discrimination is the ability to find similarities and differences among objects. Sorting and classifying are major tools for demonstrating visual discrimination. "Which of these things is not like the others?" is a game played to practice visual discrimination.

Visual memory is the ability to remember an object accurately when it is no longer visible, as when students can visualize an approximate length of an object.

Who uses spatial thinking and why should it be taught?

Many students find it easier to reason using geometry than using an analytic approach (Krutetski, 1976). Anyone who needs to visualize objects without creating the objects is likely to use spatial thinking. Architects, engineers, artists and dancers, designers, draftsman, chemists, physicists, biologists, ecologists, and mathematicians are among many who use spatial abilities. Most scientific and technical occupations require spatial abilities (Harris, 1981).

Many studies have indicated that spatial ability can be improved through teaching (Bishop, 1980). Del Grande (1987) explained ways to design geometry programs that improved students' visual perception based on his findings that it was possible to improve spatial thinking through teaching.

Spatial thinking should be taught because it has so many connections to other disciplines and to real-life situations. Guay and McDaniel (1977) suggest that mathematics achievement and spatial abilities are positively correlated. Many sorts of geometry activities that improve spatial ability can also improve success in mathematics in general (Bruni and Seidenstein, 1990). In addition, most technical and scientific occupations require persons having spatial ability at or above the 90th percentile (Clements and Battista, 1992).

Transformational geometry, mapping, reflection, glide, rotation, and dilation

Transformational geometry involves the mapping of one point to another point on a plane. Several mappings occur in transformational geometry: reflections, glides, rotations, and dilations.

A *mapping* is a projection of one point in set A to another point in set B. The counterpart to a mapping in algebra is a function. So if you were to perform the transformation shown in the following diagram, that mapped point Q from set A to point Q_1 in set B, set A would be the *domain* (the original picture) and set B would be the *range* (the new image) (Brown, 1973).

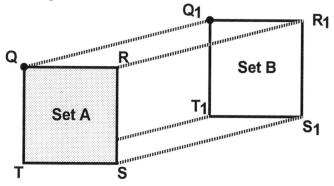

A *reflection* causes a point to be mapped across a line, across a point, or rotated about a point. A reflection is a transformation that preserves distance. By reflecting points, symmetry is created. It is termed *line symmetry* if the mapping creates a figure that creates a mirror image, as in the figure below. *Point symmetry* is when the figure can be rotated a half-turn and the image is the same as the original figure.

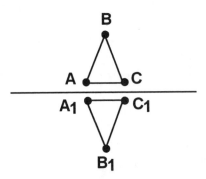

A *glide* is a translation that moves an image from one position on the page to another without any rotation of the image or reflection. A glide would be like picking up a piece of paper off your desk and putting it in the same position, two inches to the right, or moving a character on a computer screen to the left or right, up or down, without a change in size or rotation. Glides or translations preserve the size of the original image.

A *rotation* is a transformation about a point P where the image is moved about a circle with center P. Rotations can be done in a clockwise or counterclockwise direction, as shown in the following diagrams. Counterclockwise is the positive direction and the clockwise direction is negative. Rotations also preserve the size of the original image.

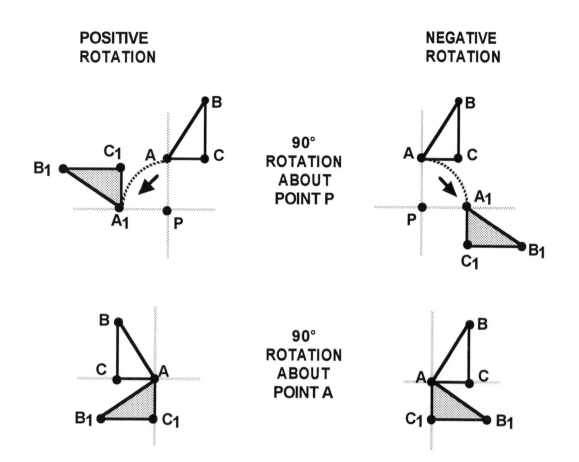

Dilation involves enlarging or shrinking an object by a scale factor, as when an image is enlarged or reduced by a photocopying machine.

C. NCTM position

K-4

Students will develop spatial sense so they might have a feeling for their surroundings and the objects in them.

5-8

Students will visualize and represent geometric figures, with special attention to developing spatial sense in one, two, and three dimensions..

Students will explore transformations of geometric figures and develop an appreciation of geometry as a means of describing the physical world.

Students will continue to study two- and three-dimensional geometry so they can interpret and draw three-dimensional objects and represent problems with geometric models.

Students will deduce the relationships between figures from given assumptions, utilizing spatial thinking.

College-intending students should be able to apply transformation, coordinates, and vectors in problem solving.

D. Integrating problem solving, reasoning, communicating, and making connections

Problem solving often requires spatial thinking, and reasoning is used to explain the relationships of the objects in spatial thinking problems. Students need to communicate spatial relationships, and activities such as those here encourage them to visualize and mathematically describe these relationships. Spatial thinking is connected to many real-life situations involving two- and three-dimensional aspects.

E. Prerequisite skills

It is important to develop positional language (*e.g.*, above, below, between, top, bottom, in front of, behind, back, side, next to) and common geometric shape names at the beginning stages of spatial thinking. Preschool age children are able to think spatially and communicate the relationships if given a chance. At this level, most students focus on the whole as opposed to the parts. At more advanced levels, students need to have an understanding of angles and their measures. Two- and three-dimensional geometry must also be understood to describe many spatial relationships.

F. Students' difficulties, confusions, and misconceptions

Students can have naive notions of the whole or the parts. Young students fail to take in the whole shape and consequently may not be able to recognize the parts.

Language and the importance of meaning in general use versus mathematical use can occasionally cause problems. If students do not understand the precision of language use in mathematics, the words might cause confusion. A teaching strategy in which the teacher explicitly explores existing meanings students have for words is important to helping them develop the desired mathematical meanings.

G. Factors contributing to students' difficulties, confusion, and misconceptions

Many difficulties are thought to be caused by developmental levels. Some of these developmental difficulties are due to a lack of exposure to spatial thinking.

Students are not encouraged to realize the importance of appropriate use of mathematical language in describing spatial relationships.

Teachers and teaching materials often do not take into account student preconceptions.

Inaccuracies can be encouraged by poor use of terminology.

Concepts are not always developed at the appropriate time and in appropriate ways to facilitate student understanding.

H. Appropriate teaching strategies

Discussion and demonstration, teaching for conceptual change, and mental model building are appropriate strategies for this topic. It is important to use discussion and demonstration so students have the opportunity to use positional language in context, as in the "What am I?" activity. Sorting and classifying activities also encourage students to explain their reasoning. "Story Paths" encourages building mental models of sequencing. "Covering cubes" also requires students to visualize models as they explore their ideas. The conceptual change strategy can be used in "Paper miniature golf," challenging students to explore the transformational geometry aspect of getting a hole-in-one.

I. Teaching notes

Words such as above, below, between, top, bottom, in front of, behind, back, side, beside, and next to are important to introduce in the primary grades. The "What am I?" activity is designed to emphasize positional words, but use of the words should not be limited to this activity. Positional words should be included as part of everyday classroom use.

The goal of the "Story paths" activity is to encourage student practice in visualizing sequences, creating paths, and representing these paths with diagrams. The motor skills needed to create straight paths are important to develop.

Use a variety of three- and two-dimensional shapes for the "Sorting and classifying" activities. For example, Geoblocks can be used to do the activity one day and attribute blocks the next.

"Paper miniature golf" uses reflections of points across lines to solve the problem of where to hit a golf ball to get a hole-in-one. Graph paper or a Mira board can be used. Paper folding also can be used to reflect the point across the line. Fold the paper on the "wall" (line) and mark the point to be reflected. Challenge the students to bank their shots off two, three, and even more walls.

J.   Materials needed

> appropriate stories (ones in which the characters go from one place to another)
> yarn
> diagrams for stories (these will vary for each story)
> Geoblocks
> attribute blocks
> pattern blocks
> items to sort and classify
> cubes and corresponding-sized grid paper (1-inch cube, 1-inch grid paper)
> scissors
> Mira boards
> graph or grid paper
> rulers, protractors, compasses

K. Activities

See the following pages.

1. What am I?

This activity is not intended as a worksheet. It is designed to be used as discussion and demonstration with the early primary class.

Choose a visible item in the classroom. Using positional terms such as above, below, between, top, bottom, in front of, behind, back, side, and next to describe the location of the item without saying what it is. Ask students to visually explore the room and write down the item on their math pad when they have figured it out. For example, you might choose the stapler on top of the desk. You could say, "I'm on top of a desk. The desk is next to the window with a chalkboard in front, and another desk behind." Continue to eliminate other possible items by including more clues.

A reverse of this game is to choose an item and have students describe the location using appropriate positional language. After a few practices at this, children can partner and play the game in pairs, where one describes and one figures out the object.

2. Story paths[5]

This activity is not intended as a worksheet. It is designed to be used as discussion and demonstration with the early primary class.

Read a story such as "Goldilocks and the Three Bears." Act it out with the children, making sure to show the path that the characters take. Keep the paths obvious by using two different colors of yarn to follow the characters along. After the story, talk about the path of the yarn in relation to the story. Have the children remember the order of the story using the yarn as a guide.

Another day, as you reread the story, have the children draw a path representing Goldilocks travels through the house on a prepared house diagram. After the paths are complete, have the children tell each other the story based on their diagrams.

---

[5] Adapted from: Bruni, J. V. & Seidenstein, R.B. (1990). Geometric concepts and spatial sense. In J. N. Payne (Ed.). *Mathematics for the Young Child*. Reston, VA: NCTM.

3. Sorting and classifying

This activity is designed for discussion and demonstration with the early primary class.

Be sure each child has attribute materials or other materials that have shapes to work with.

a.   Alike and different

Sort the items by shape. Put the like shapes in piles. Describe why those shapes are alike. Describe why those shapes are different.

b.   One of these is not like the others

Find shapes that will show this: "One of these shapes is not like the other shapes." Have your partner identify the shape that is not like the others.

c.   Defining attributes

Discuss "What makes a square different from a triangle?"

d.   Two-pile sorts

Ask the children to show how they would sort the circles, triangles, and squares into only two piles and explain their reasons for sorting them as they do.

The children may choose "shapes with edges" (or corners) or "no edges" (or corners). Use the children's language and if the terms are not as accurate as necessary, find examples that make this apparent to encourage the children to use words in a meaningful manner.

e.   Trains

Have students create trains with their shapes where each piece is a different shape from those next to it. Add two-difference, then three-difference shapes.

f.   Venn diagrams

Use Venn diagrams to sort and classify the shapes. Have students create labels to describe their sorts.

4. Covering Cubes

How many different shapes of connected squares will cover the surface of a cube? For example, if I cut out the shape below, would it cover the cube so that no face was left uncovered?

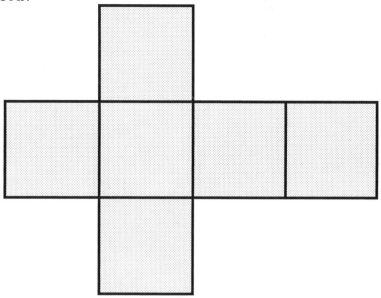

a.

Predict how many different shapes of six connected squares will cover the cube.

Then cut out shapes from the grid paper that will completely cover a cube.

Test your ideas by folding the shapes into cubes.

Compare your findings with your partner(s).

Have you found all of the shapes possible?

b.

Predict what shapes cannot be folded to create a cube from six connected squares.

Cut them out and show why they do not work.

Create two piles: one pile of shapes that will cover a cube and one pile of shapes that will not cover a cube. Which pile is bigger?

How do your results compare to your predictions?

5. Paper miniature golf[6]

Reflections of points across walls of an imagined miniature golf course can be used to solve the problem of how to hit a ball to score a hole-in-one.

Example:

Consider the hole layout depicted in Figure 1, below.

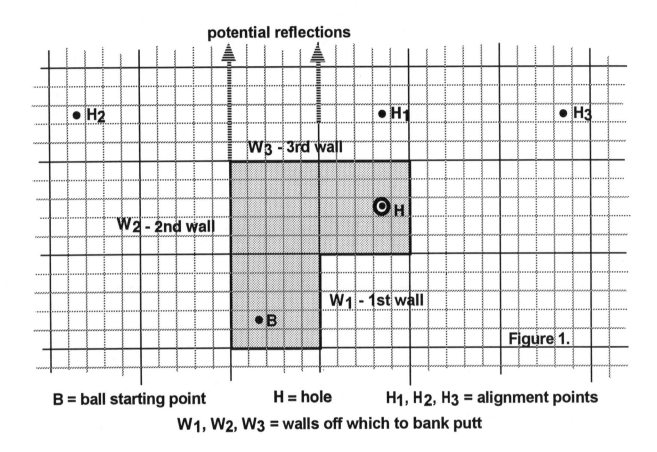

B = ball starting point   H = hole H_1, H_2, H_3 = alignment points
W_1, W_2, W_3 = walls off which to bank putt

Here is how we would figure out how to hit a <u>hole-in-one</u> for this hole; that is, hitting the golf ball from starting point B in such a way that it will fall into the hole (H) with a single stroke from the club.

1.   First reflect the center of the hole (H) across the last wall you want to hit (W_3). It is easy to do this by aligning a Mira board on W_3 and marking the reflection of H. In the figure, the resulting point is H_1. Note that H_1 is the same distance from the wall as H and on the same line.

---

[6] Adapted from: Powell, N.N., Anderson, M., and Winterroth, S. (1994). Reflections on miniature golf. *Mathematics Teacher, 87,* 490-497.

---

2.   Next, reflect H$_1$ across the second wall (W$_2$) you want to hit with the ball, resulting in point H$_2$.

3.   Then reflect H$_2$ across the first wall you want to hit (W$_1$) to obtain point H$_3$.

4.   Now that you have marked where the ball should be aimed, draw the lines as shown. Connect the ball to the last reflection point, H$_3$. The result will be as in Figure 2, below.

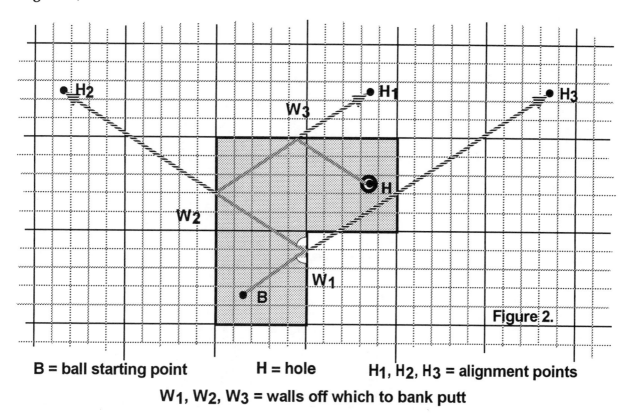

B = ball starting point   H = hole   H$_1$, H$_2$, H$_3$ = alignment points
W$_1$, W$_2$, W$_3$ = walls off which to bank putt

The alignments shown in Figure 2 show the path the ball should travel for a hole-in-one.

Check your work by assuring yourself (and your teacher) that <u>the angle at which the ball hits the wall and the angle it leaves to travel to the next destination are congruent</u>. The angles should be equal in measure, as shown.

The specific instructions given are for a ball that would hit the wall three times (a three-bank shot). The same directions will work, however, for other numbers of banks in a single shot. Also, you should know that not all miniature golf holes have a one-hole solution because of the design of the holes.

a.

Demonstrate how to get a hole-in-one for the following golf holes (1-3).

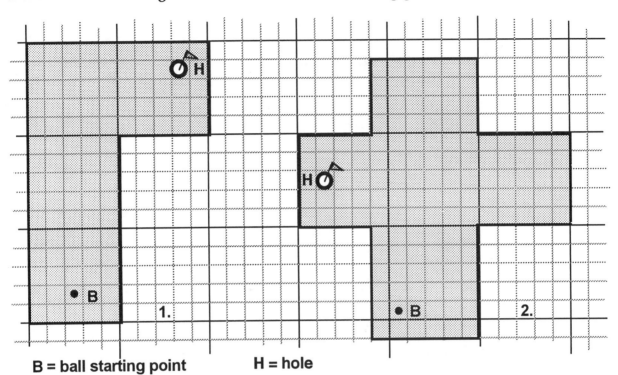

B = ball starting point   **H = hole**

B = ball starting point   **H = hole**

b.

Using grid paper, create a 9-hole miniature golf course.

Remember that not all of the designs for a hole can result in a hole-in-one.

For each hole, write down the *par* = the minimum number of shots it would take to get the ball into the hole.

c.

Exchange your golf course with a partner.

Try to complete each other's course with the minimum number of shots.

L. Assessment ideas

For performance assessments, use pattern blocks or attribute blocks to recreate the shapes in a picture; continue a pattern made with shapes; transform letters using rotations, slides, and flips; and use Geoboards to make congruent shapes.

Have students describe the positions of different shapes in drawings in their journals.

Use maze tracings that include finding the shortest path to a goal and identifying non-conforming transformations in pictures.

Students can demonstrate their understanding of transformational geometry by creating "Escher drawings." They also could use such computer software as The Factory™, SuperFactory™, Geometer's Sketchpad™, Cabri™, and others for spatial thinking projects.

M. Resources and references

Bishop, A. J. (1980). Spatial abilities and mathematics achievement--A review. *Educational Studies in Mathematics, 11,* 257-269.

Brown, R. (1973). *Transformational Geometry.* Palo Alto: Dale Seymour.

Bruni, J. V. & Seidenstein, R. B. (1990). Geometric concepts and spatial sense. In J. N. Payne (Ed.) *Mathematics for the Young Child.* Reston, VA: NCTM.

Clements, D. H. & Battista, M. T. (1992). Geometry and spatial reasoning. In D. A. Grouws (Ed.) *Handbook of Research on Mathematics Teaching and Learning.* Reston, VA: NCTM.

Del Grande (1987). Spatial perception and primary geometry. In M. M. Lindquist (Ed.) *Learning and Teaching Geometry, K-12: 1987 Yearbook.* Reston, VA: NCTM.

Guay, R. B. & McDaniel, E. D. (1977). The relationship between mathematics achievement and spatial abilities among elementary school children. *Journal for Research in Mathematics Education, 8,* 211-215.

Harris, L. J. (1981). Sex-related variations in spatial skill. In L. S. Liben, A. H. Patterson, & N. Newcombe (Eds.), *Spatial representation and behavior across the life span* (pp. 83-125). New York: Academic Press.

Krutetski, V. A. (1976). *The Psychology of Mathematical Abilities in Schoolchildren.* Chicago: University of Chicago.

National Council of Teachers of Mathematics. (1989). *Curriculum and Evaluation*

Standards for School Mathematics. Reston VA: Author.

Powell, N.N., Anderson, M. and Winterroth, S. (1994). Reflections on miniature golf. *Mathematics Teacher, 87,* 490-497.

13
ALGEBRA

A. Overview

Concepts included in this chapter are: uses and importance of algebra, language of algebra, connecting data to graph to statement to symbols, types of relationships, and equations.

B. Background for the teacher

Ask an engineer, a scientist, a banker, or an economist, and each will tell you how she uses algebra. In addition, a carpenter, a dietitian, and a tax person all rely heavily on algebra. People in many other occupations unknowingly use algebra in their businesses or at home.

The language of algebra

Algebra uses terms like equation, variable, function, linear, inverse, and quadratics. An *equation* is a mathematical sentence that expresses how one quantity is related to another. For example, the equation $F = 9/5\ C + 32$ is a mathematical sentence expressing how F and C are related.

Equations contain both known information and unknown information. A *variable* is an unknown quantity we want to find. The equation $F = 9/5\ C + 32$ is a formula for converting temperatures from Celsius (C) to Fahrenheit (F). In this equation, C is the *independent variable* and F is the *dependent variable*. The value of the dependent variable <u>depends on</u> the value of the independent variable.

A *function* is a relation between two variables where for each value of independent variable there is only one value of dependent variable.

What type of algebraic relationships do we use?

The relationship of F to C above is a *linear* relationship. In a linear relationship, F changes directly and proportionally as C changes. Any general linear relationship that relates variables y and x may be expressed as $y = mx + b$. In this relationship, **m** (the *coefficient* of the independent variable) is the *slope* and **b** is the *intercept*.

For a rectangle of given length, if we increase the length, the area will increase proportionally. For example, for a rectangle 2 cm wide and 3 cm long, the area is 6 cm^2. If we double the length (l), keeping the width constant, the area (A) also doubles; therefore, we say the area of a rectangle of a given width varies linearly with length. The equation for this example is $A = 2l$.

In an *inverse* relationship, one variable decreases as the other increases. For example, if we keep the temperature of a gas constant and increase the volume, the pressure will decrease. Here we say volume and pressure have an inverse relationship. This we represent as $PV = C$. P is the pressure, V is the volume, and C is a constant. Pressure (P) is measured in F per unit area. Some common units for pressure are lb/in^2 and N/m^2. (N represents the unit newton.)

Quadratic relationships involve equations in two variables, with at least one variable of degree two but none higher than degree two. For example, the area of a square of side 2 cm is 4 cm^2. If we double the side to 4 cm, the area is 16 cm^2. The area varies as the <u>square</u> of the side, so there is a <u>quadratic</u> relationship between the area of the square and its side. Examples of quadratic relationships include the parabola, circle, hyperbola, and ellipse. These are called the *conic sections*.

What are the steps in writing a mathematical equation?

Given the mathematical sentence, the student first should understand what the words in the problem represent.

The second step is translating the situation (the problem) into the student's own words or meaning.

Third, identify the questions:

> What do we want to find?
> What is given?
> What does each variable represent?

Fourth, choose variables (letters) to represent quantities.

Fifth, present ways to isolate the variable in question.

C. NCTM position

Students will learn the concepts of variable, expression, and equation.

Students will learn how to represent situations and number patterns with tables, graphs, and equations and explore the relationships of these representations.

Students will analyze tables and graphs to identify properties and relationships.

Students will develop confidence in solving linear equations using concrete, informal, and formal methods.

Students will apply algebraic methods to solve a variety of real-world and mathematical problems.

Students will learn about different relationships.

D. Integrating problem solving, communicating, reasoning, and making connections

As students are involved in <u>doing</u> algebra, they will be using problem-solving strategies and reasoning. Many activities in the algebra unit connect meaningfully by drawing on science concepts and situations from students' examples. Each activity encourages students to use graphs, equations, and statements to communicate relationships.

E. Prerequisite skills and knowledge

To be successful in algebra, the student should have basic operations skills and knowledge of decimals and fractions. In addition, the student needs knowledge of basic geometry, including area and volume of simple shapes and appropriate units. The student also should have the skill of translating mathematical ideas from the concrete to abstract.

F. Students' difficulties, confusion, and misconceptions

Extensive research by mathematics educators from many countries reveals that a large number of students entering algebra classes have difficulty and misconceptions in the area of variables. They also have difficulty distinguishing between constant and variable. The notion of function also causes problems.

G. Factors contributing to students' difficulties, confusion, and misconceptions

Among the factors contributing to students' problems with algebra are:

- the abstract nature of algebra, given that many students are concrete thinkers;
- the sequence of presentation of algebraic concepts, from abstract to concrete;

- the emphasis in textbooks and teaching on the symbolic portion of algebra combined with a lack of connection of algebraic symbolic presentation to real situations that are familiar to students;
- assignments that emphasize repetitively solving equations rather than applying algebraic thinking to practical situations;
- students' inadequate understanding of necessary prerequisite concepts.

H.   Appropriate teaching strategies

Teaching for conceptual change, discussion and demonstration, and concept mapping are useful strategies for teaching algebraic concepts. Concrete examples and manipulatives also should be used.

I.   Teaching notes

In the spirit of what we have been proposing throughout the book (helping students to DO mathematics) we suggest that teachers challenge students to DO algebra. Begin with the concrete and slowly move to the abstract. Give students the opportunity to collect their own data, use their own symbols, and write sentences in their own words. Have students connect what is being discussed in class to other situations.

For activities 7, 8, and 9 provide appropriate boxes made from graph or grid paper. For activity 7 provide students with boxes of constant height and different size base areas. In activity 8, students should have access to boxes of the same base and different heights. Activity 9 calls for boxes of different size square bases and constant height.

To prepare students for activities 7, 8, and 9, you may use the examples below, which should help the students feel more comfortable connecting algebra and geometry.

Example:

A rectangle of dimensions 5 cm x 3 cm has an area of 15 cm2. A rectangle of dimensions 7 cm x 3 cm has an area of 21 cm2.

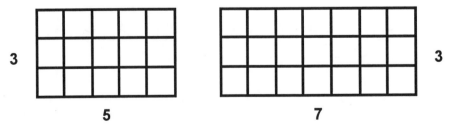

The plot of area against length using a table of data similar to the one below for rectangles of constant width, such as 3 cm, yield a straight line of slope 3 cm.

Area (cm²)	Length (cm)
15	5
21	7
24	8
30	10

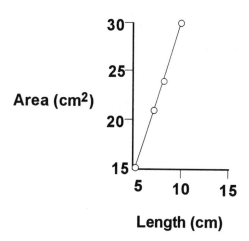

J. Materials

liquids: water, vegetable oil, syrup, and saturated salt solution
food coloring
500 ml and 1000 ml graduated cylinders
100 and 250 ml beakers
thermometers with Fahrenheit and Celsius scales
string
pendulum bobs
balls
Meter sticks
washers or weights
fulcrum
syringes
materials of various sizes, to include wood, plastic, minerals, and beans
cylinders of various sizes (made of any kind of material)
rectangular prisms, cubes, and cones of various sizes
balances and weights
spring scales
grid and/or graph paper

K. Activities

See the following pages.

1. Relationship of mass and volume of liquid

You are presented with a challenge of determining the relationship between the mass (in grams) and the volume (in ml) of a given liquid.

Take at least 4 readings, ranging from about 20 ml to about 100 ml. Construct a table like the one below and determine the ratio of mass to the volume of the liquid.

M (gm)	V (ml)

Graph the data.

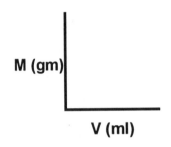

M (gm)

V (ml)

Using the table or the graph, make a statement about the relationship between the mass and the volume of the liquid.

Choose symbols for the variables and write an equation describing the relationship between the mass and volume of the liquid.

Using the table, the graph, the statement, or the equation, predict

      (a) the mass of a small volume of liquid (not measured),

      (b) the mass of a large volume (not measured).

How would you test your prediction?

What is the slope of the graph?

2. Fahrenheit (F) versus Celsius (C)

Imagine you are in your hotel on vacation in Southeast Asia. You feel you are coming down with a severe fever. How would you find your body temperature? Looking around, you find a fever thermometer that is marked in degrees Celsius. The thermometer registers your temperature as 40° C. Should you be concerned? You understand the meaning of Fahrenheit readings of body temperature, but don't know how to interpret the Celsius readings. Then you find a partial chart like the one below, showing some Fahrenheit measurements and the corresponding degrees in Celsius.

F°	C°
212	100
158	70
50	10
-4	-20

Using this information, how would you figure out what your temperature is in degrees Fahrenheit?

Make a graph comparing F to C.

F°

C°

Write a statement about the relationship between F and C, describing as <u>much</u> as you can about how the scales are related.

Write an equation describing the relationship between the scales.

Using the graph or the equation
    (a) predict different values of F for a given C
    (b) predict different values of C for a given F

Using thermometers, check your predictions.

What is the slope of F vs C? What is the slope of C vs F?

How do we represent C = 0° on the graph? How do we represent F = 0° on the graph?

What is the Fahrenheit equivalent to your body temperature of 40° C? Should you be concerned?

---

3. Prices of candy and gum

Two packs of a particular gum and a candy bar cost 76¢. Two candy bars of the same brand and one pack of the same brand cost 80¢. You need to find the answers to two questions: "What is the price of a pack of gum?" and "What is the price of a candy bar?"

In your group decide on a plan to solve the problem.

Share your plan with others.

Implement the plan and share the results with others.

How can you test your answer?

Present examples similar to the gum and candy problem that can be solved using your approach.

4. Balancing

Predict which of the following three situations will balance.

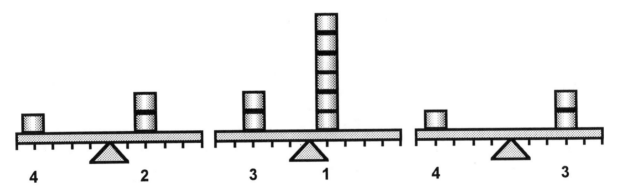

Test your predictions by working with the materials.

Write a statement describing the relationship between weight and the distance from the fulcrum in equilibrium situations.

5. Which things will float on water and which will sink?

You will be provided with objects of varying sizes and made of various materials; *e.g.*, wood, plastic, beans, and minerals.

Collect and tabulate the <u>mass</u> and corresponding <u>volume</u> for each object (take at least 4 readings for each object).

Based on your data, graph mass versus volume *for each material.*

Name of Object:

Mass	Volume

What is the slope of each graph? What is the physical meaning of the slope?

Knowing that the slope of the graph of mass against volume for <u>water</u> is 1 gm/ml, which of the solids do you think will float on water? Which do you think will sink?

6. Pressure and volume

In the situation depicted below, predict the changes in volume that may occur in a syringe as you change the pressure.

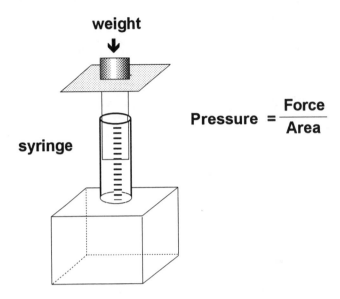

$$\text{Pressure} = \frac{\text{Force}}{\text{Area}}$$

Test your ideas by working with materials and collecting appropriate data.

Write a statement describing the relationship between volume and pressure as revealed by your experiments.

Present the relationship symbolically.

What are some other examples of this phenomenon?

7. Volume versus area of rectangular solids of the same height

If you were to plot the volumes of rectangular solids (prisms) provided by your teacher against their areas of the base, what graph would result?

Write your prediction and explain your reasons.

Share your prediction with others in your group.

Have someone present all of the predictions and explanations of the members of your group to the class.

How can your test your predictions?

Test your ideas, record your observations, then present them as a graph.

Based on your observations, what changes would you like to make in your original explanation for what would happen?

What is the unit of the ratio of volume to the base area of rectangular solids (prisms)?

How can you symbolically represent the relationship you have discovered?

What would happen if you used cylinders instead of prisms? Explain your answer.

What would happen if you used cones? Explain your answer.

8. Volume versus height of rectangular solids of the same base

What relationship do you think would result if you plotted the volume against the height of rectangular prisms (boxes)?

Write your prediction and explain your reasons.

Share your prediction with others in your group.

Have someone present all of the predictions and explanations of the members of your group to the class.

How can you test your predictions?

Test your ideas, record your observations, then present them as a graph.

Based on your graph, what changes do you want to make in your original explanation?

What is the unit of the ratio of volume to the height of rectangular prisms?

Represent this relationship symbolically, defining each symbol.

9. Volume versus side of the base of a rectangular solid having a square base and constant height

What relationship would result if you plotted the volume of a rectangular solid having a square base against the side of the base?

Write your prediction and explain your reasons.

Share your prediction with others in your group.

Have someone present all of the predictions and explanations of the members of your group to the class.

What do you need to test your predictions?

Test your ideas, record your observations, then present them as a graph.

Based on your findings, what, if any, changes do you want to make in your explanations?

What statement can you make about the relationship between the volume of a solid and the length of the base?

What are the units (both Metric and English) that are involved here?

What would happen to the relationship if you plotted the volume against the radius of the base of cylinders? of cones? Explain your answers.

10. Liquids and solids

You are given three different liquids and three different solids.

In your group, decide what steps you will take to:

    (a) determine what will happen if you pour the liquids in each other, and

    (b) determine what will happen to each solid as it is placed in the container containing each liquid.

Get the necessary materials and pursue your plan.

Construct appropriate tables of data.

Construct appropriate graphs.

What statements can you make about your findings?

11. Pendulum

Work in small groups to make and explain predictions, test your predictions, collect data, organize and interpret the data, and draw conclusions.

Predict how the *period* of a pendulum (the time it takes for the pendulum to make a complete swing) varies with the *effective length* (distance from the point of suspension to the center of mass).

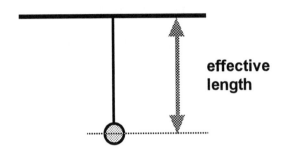

Construct a table to record data, then test your prediction using the pendulum apparatus.

Write a statement describing the relationship between the effective length of the pendulum and its period.

Write a general algebraic expression describing the relationship.

Test the expression by collecting additional data.

12. Dropping a ball (height versus time)

If you drop a ball, what do you predict is the <u>relationship</u> between the height from which it is dropped and the time it takes to fall to the ground?

To test your prediction of the relationship, examine the table of data showing the heights from which an object is released and the time it takes for it to reach the floor.

Height (m)	Time (sec)
5	1
20	2
45	3
?	6
?	10

Write a statement describing the relationship between distance (m) and time (sec) revealed in the table.

How would you represent the relationship symbolically?

What would be the height of the drop if the ball reached the ground in 6 seconds?

If it took 10 seconds?

How long would it take the ball to drop from a height of 225 m?

13. Concept mapping "algebra"

Write down as many terms as come to mind about the word <u>algebra</u>.

Then write down connectors that could be used to connect these terms to the idea of algebra <u>and</u> to each other

In your group, write sentences using the terms and connectors you have thought about.

In your group, design a concept map.

Share your map with other groups.

As others present their maps, make appropriate modifications to your concept map.

Consult books and other sources to further revise your map.

Identify questions and problems on algebra which stemmed from the concept maps.

14. Bring examples to share

Identify examples from sports, cooking, cars, sales, and other areas of how linear equations, inverse relations, and quadratic equations are used.

L. Assessment strategies

For this unit, various assessment strategies can be used to determine the degree to which the students have met the expectations. These strategies include teacher observations, student journals, student projects, concept mapping, pencil-and-paper tests on student understanding of algebraic concepts, and interviews.

M. Resources and references

Center for Occupational Research and Development. (1988). *Solving Problems that Involve Linear Equations.* Waco, TX: Author.

Ebert, C. L. (1993). *An Assessment of Prospective Secondary Teachers' Pedagogical Content Knowledge about Functions and Graphs.* Paper presented at the annual meeting of AERA, Atlanta, GA.

Goodson-Espy, T. J. (1995). A constructivist explanation of the transition from arithmetic to algebra: the role of reflective abstraction. *ERIC Reports-Research/Technical (143).* Washington, DC: U. S. Department of Education.

Kaur, B., Peng, S., & Boey, H. (1994). Algebraic misconceptions of first year college students. *Focus on Learning Problems in Mathematics, 16*(4), 43-58.

Mevarech, Z., & Kramarsky, B. (1993). How Often and Under What Conditions Misconceptions Are Developed: The Case of Linear Graphs. *Proceedings of the Third International Seminar on Misconceptions and Educational Strategies in Science and Mathematics.* Ithaca, NY: Misconceptions Trust.

Movshovitz-hadar, N. (1993). A constructive transition from linear to quadratic functions. *School Science and Mathematics, 93*(6), 288-298.

National Council of Teachers of Mathematics. (1989). *Curriculum and Evaluation Standards for School Mathematics.* Reston VA: Author.

Stepans, J. I. (1982). Teaching mathematics using the learning cycle. *Journal of Developmental Education, 1*(2), 6-8.

Stepans, J. I. (1982). I don't know what you mean. *The American Association of Two-Year Colleges, Fall 1982,* 11-17.

Stepans, J. I.(1985). Mathematical conceptions or misconceptions of middle school students. *Transescence, 13*(1), 20-24.

Stepans, J. I. (1991, May). Will it mix, float, or sink? *School Science and Mathematics,* 218-220.

Stepans, J. I. (1995, February). The power of mathematics is in communicating, predicting, and verifying. *Illinois Council of Teachers of Mathematics,* 16-18.

Stepans, J. I. (1996). *Targeting Student's Science Misconceptions.* Riverview, FL: Idea Factory.

Stepans, J. I., & Olson, M. (1985, January). We should be teaching them more than symbolism. *School Science and Mathematics,* 1-10.

Vandeputte, C. (1993). Misconceptions in Space-Time Graphs. *Proceedings of the Third International Seminar on Misconceptions and Educational Strategies in Science and Mathematics.* Ithaca, NY: Misconceptions Trust.

14
DATA AND STATISTICS

A. Overview

Concepts included in this chapter are: history and uses of statistics; collecting, organizing, interpreting, and using data; mean, mode, and median; central tendency, normal curve, range, standard deviation; and applications of statistics.

B. Background for the teacher

Statistics involves gathering, tabulating, analyzing, and drawing inferences from data (information). Historical examples of the use of statistics are abundant. The first statistics dealt with keeping track of mortality. Augustus developed an accounting sheet of the Roman Empire. Israel was the first nation to do a census. William the Conqueror's use of statistics appeared in the 1085 *Domesday Book*. In 1532 Graunt used data from various London churches to to draw his conclusions about mortality rates.

Others who have used statistics and have made contributions to the field include Halley (famous for the comet of the same name) in the 18th Century and DeMoivre, who developed the mathematics of statistics, particularly dealing with combinations and permutations. Lagrange and Laplace, among others, improved the field and applied it to the study of celestial mechanics.

Present uses of statistics

To respond to comments such as, "What is the top movie of the week?" "What is the most popular song of this month?" "What is the average salary of a college graduate?" and "What are the standings in the NFL?" people rely on statistics. Users of statistics include insurance companies, politicians, stores, the medical profession, sports people, the Census Bureau, and quality control people, whether they are examining the quality of a product, air, or water. With the skyrocketing popularity of the Internet, data--both accurate and inaccurate--is more available than ever.

Information is needed as the basis for making intelligent decisions, yet raw information is often inadequate or misleading. Statistics--the science of gathering, tabulating, and analyzing data--is crucial to decision making.

Some terms associated with statistics

The tendency of numbers in a set to cluster around the middle of a set is called *central tendency*. Measures of central tendency include mean, median, and mode.

The *mean* is the sum of all the values in a set divided by the number of values in the set. Most people use the terms mean and average synonymously, but average does not imply just mean. A median and mode are also ways to find an average.

The *median* is the value in the middle of a data set when the data are arranged in numerical order. Or, if there are an even number of values in the set, it is the mean of the two middle values.

The *mode* is the value in the data set that occurs most frequently. There can be more than one mode. A set of data with two modes is called bimodal.

A *histogram* is a graph of frequency distribution. It is a critical graph in statistics for showing frequency.

A *normal curve* is the graph of a hypothetical, ideal frequency distribution, having the same values for mean, median, and mode.

The *standard deviation* is one common indicator of variability in a set of data. The figure that follows illustrates standard deviation. In a normal curve, 68% of the data falls in a range one standard deviation in both directions from the mean.

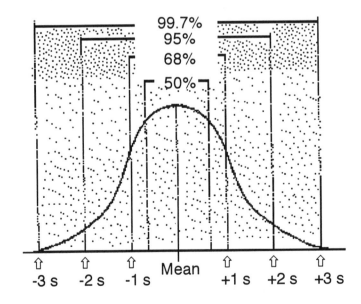

How do we collect, organize, and use data?

Data (information) can be collected in a variety of ways, including conducting interviews and surveys, looking at test scores, reviewing files and portfolios, and making observations. Data (information) may be organized in a table or various kinds of graphs; *e.g.*, bar, line, and pie graphs. Tables and graphs make it possible for us to identify and study relationships, if there are any.

Organized data, whether in the form of a table, graph, or statements, enable us to go beyond the immediate facts. Interpolating and extrapolating are among many statistical techniques that allow us to make powerful decisions. *Interpolate* means to insert estimated values between two known values. *Extrapolate* means to project, extend, or make inferences from a trend that is apparent in a known data set.

C. NCTM position

K-4

Students will collect, organize, and describe data.

Students will construct, read, and interpret displays of data.

Students will formulate and solve problems that involve collecting and analyzing data.

5-8

Students will systematically collect, organize, and describe data.

Students will construct, read, and interpret tables, charts, and graphs.

Students will evaluate arguments that are based on data analysis.

Students will be able to make inferences and convincing arguments that are based on data analysis.

Students will develop an appreciation for statistical methods as a powerful means for decision making.

9-12

Students will construct and draw inferences from charts, tables, and graphs that summarize data from real-world situations.

Students will use curve fitting to predict from data.

Students will understand and apply measures of central tendency, variability, and correlation.

Students will understand sampling and recognize its role in statistical claims.

Students will be able to design a statistical experiment to study a problem, conduct the experiment, and interpret and communicate the outcomes.

Students will analyze the effects of data transformations on measures of central tendency and variability.

D. Integrating problem solving, reasoning, communicating, and making connections

In the activities on statistics, every effort is made to encourage students to verbalize the problem (question), propose appropriate strategies they will use to find solutions to the problem, and implement appropriate operations. In addition, the students are given activities and questions in which they connect statistics to everyday situations. As they attempt to find solutions to the problems posed in the activities, the students are encouraged to use reasoning and to use mathematics to communicate their findings and ideas.

E. Prerequisite skills and knowledge

In activities targeted for secondary students, the students should already be able to use tables and graphs, use formulas, and solve linear equations.

F. Students' difficulties, confusion, and misconceptions

Students think that any difference in the means between two groups is significant.

Many students believe that there is no variability in the world.

Many students draw generalizations from data based on small samples.

Many students do not appreciate small differences in large samples.

Most students believe that the size of a random sample is independent of the size of the overall population.

Some students draw generalizations without considering the sample size.

Students estimate likelihood for events on how well an outcome represents some aspects of its parent population.

Some students have a tendency to perceive each piece of data as a separate, individual phenomenon.

Some students feel that even smaller sample sizes are sufficient in a replication study.

Many students have difficulty with the concepts of mean and variance.

Many students are confused about the sample mean in situations where it must be calculated as a weighted average.

Many students fail to understand that doing statistics involves drawing an appropriate sample from a defined population, not sampling the whole population.

Many students do not know how to choose between the mean, median, and mode as an appropriate method in a given situation.

G. Factors contributing to students' difficulties, confusion, and misconceptions

Most K-12 schools devote a small amount of time on the topic of statistics.

Many students do not have the opportunity to improve their statistical intuition because schools do not provide adequate experiences.

Lack of preparation of teachers contributes to students' inability to have appropriate experiences in statistics.

Students are not given the opportunity to confront their beliefs about statistics.

Students learn and practice a series of separate mathematics skills. Using these, they may solve unrealistic problems.

H. Appropriate teaching strategies

Appropriate teaching strategies include teaching for conceptual change, mental model building, collaborative learning, and discussion and demonstration. For this chapter we suggest using the conceptual change strategy with activities 6 and 7. Activities 1, 2, 3, 4, 9, 10, and 11 lend themselves well to collaborative learning. Activities 5, 8, and 12 may be approached with class discussion.

I.   Teaching notes

In this chapter we have limited the mathematics to basic operations and simple algebra. Consider using the Internet as an information source for basic operations for census data, insurance information on seat belts, consumer information, weather, or other data sets.

J.   Materials

> census bureau charts on United States population
> color samples for Activity 3
> scientific calculator
> graphing calculator
> graph paper
> newspaper clippings on batting and pitching averages for baseball players
> price lists of pizzas from various restaurants
> information of connection between drugs and crime
> information/studies on impact of wearing seat belts on injuries during
> > accidents

K.   Activities

See the following pages.

1. School sports

Your school needs to make decisions about the most popular sport for girls and the most popular sport for boys. The sports discussed are basketball, football, volleyball, swimming, skiing, track, and baseball. Your class has volunteered to help the school with collecting and tabulating the needed data.

In your group review the problem.

> What is given?
>
> What is to be found?
>
> What are the parameters; *e.g.*, the length of time, sample size, etc.?

In your group decide what data you will need.

Decide on how you will collect the data.

Decide on how you will tabulate the data.

How will you present the data?

How will you make decisions based on the data?

Carry out the plan.

Present your results to other groups.

What are the limitations of your study?

As a class, decide on the most effective study. What criteria were used to decide?

What is the most effective way to present the data?

How did you choose your graph type? your sample size?

2. What is happening to the population?

The Census Bureau needs your help in determining how the population of the United States is growing.

The challenge for your group is to present a model for population growth for the past 100 years. Then, based on the trend, can one predict what the population will be in 50 years?

In your group decide what is the problem to be studied: What is to be found?

What data will you collect?

Where will you get the information you need?

How will you tabulate the information?

Collect the necessary data.

Tabulate the data, using tables, charts, etc.

Does the population growth follow a recognizable model?

Identify some of the events which have affected any change in the trend.

Based on the model you have constructed, can you make predictions on the status of change in population of the United States?

Why or why not?

3. Favorite color combination for girls' clothes

A local merchant needs your help in deciding what color outfits she should order for her store. What is the most popular color combination of girls' shirts and slacks?

In your group review the problem.

What information will you need to solve the problem?

How will you go about collecting the data?

Will you use the total population? Why or why not?

How will you tabulate the data?

Collect the data and tabulate your data.

What conclusions can you draw?

Prepare to present what you found to the other groups in class.

What limitations did you identify in your study?

4. Favorite name for a boy? for a girl?

The neighbors are expecting twins. They have been told that the babies will be a boy and a girl. They want to give their babies the most popular names. You can help.

In your group review the problem.

Outline your plan.

How will you go about collecting the data?

How will you summarize the data?

Collect your data and tabulate the data.

What are your conclusions?

How consistent are your findings with those nationally?

Prepare to present what you found to the other groups in class.

What limitations did you identify in your study?

5. Is it a normal distribution?

Some teachers like to grade on a curve. Here is your chance to provide advice for these teachers.

A teacher just finished grading an examination. Here is the distribution of grades for his class of 22 students:

76	45	67	92	77	90	69	87	58	63	72
78	88	65	76	94	59	60	87	49	70	76

Help the teacher with the following questions:

What was the average (mean) grade?

What is the mode?

What is the median?

The teacher is considering assigning grades according to standard deviation, in which

  a. scores more than two standard deviations above the mean = A;

  b. scores between one and two standard deviations above the mean = B;

  c. scores between the mean and one standard deviation above the mean = C;

  d. the rest of the scores will = D.

How many students will receive A's? B's? C's? D's?

Is this a normal distribution? How could you tell?

6. Is a good pitcher necessarily an excellent batter?

Each of you select a favorite professional baseball pitcher who is also a hitter. Argue the fact that he is or is not also an excellent batter.

To test your idea, propose how you will go about collecting your data.

For one or two weeks, keep track of the pitching data you have chosen to collect for your favorite pitcher. Also keep track of his batting average.

What are his standings as a pitcher?

What are his standings as a batter?

What statements can you make about your favorite pitcher?

7. Is the fastest pitcher necessarily the best pitcher?

What is your prediction in regard to this question: "Is the fastest pitcher in a given league necessarily the best pitcher?"

How did you define "best pitcher?"

Share your predictions with others in your group.

As a group, decide how you will go about collecting the information you need.

Collect the necessary data and share your results with others, using appropriate tables or charts.

What statements can you make about your study?

8. You create a problem for others

Each group will pose a problem that deals with statistics and challenge the others to solve it. The problem should be approved by the instructor.

9. Best buy

You have a choice of several pizza places. As a group, you want to decide what is the best buy.

In your group, decide what the concept of "best buy" means to you.

Decide on a way to collect the necessary information.

Collect the information.

Tabulate the information and draw conclusions.

Present your data and conclusions to other groups.

10. Is there a connection between crime and drugs?

Let's see if we can answer this question.

In your group, discuss the issue. What is the problem to be studied?

What plan do you have to pursue the issue?

What data do you need to collect?

Where will you get the information?

How will you tabulate the information?

Who will be involved in data collections?

Carry out your plan.

Prepare a presentation for the class.

What were some of the limitations of your study?

11. Seat belts and accidents

We read a great deal about seat belts saving lives. Do a quick survey of the class.

We want to find out two things:

  a. Is this class a representative sample? and
  b. Is there truth to the statement "Seat belts save lives."

In your group, determine:

  what information is needed to respond to (a), and
  what information is needed to respond to (b).

Design a plan

Where will you get your data?

How will you organize the data?

Carry out the plan.

Organize the data.

Prepare a presentation for the class.

What are some of the limitations of your study?

What suggestions do you have for others who may want to conduct a similar study?

12. Pick your own project from the list

Here are some projects to study:

    traffic
    comparing costs
    politics
    weather
    agriculture
    most popular movie
    salaries versus years of education

Decide what information is needed.

Design a plan.

Where will you get your data?

How will you organize the data?

Carry out the plan.

Organize the data.

Prepare a presentation for the class.

What are some of the limitations of your study?

What suggestions do you have for others who may want to conduct a similar study?

L. Assessment ideas

This chapter lends itself to students collaborating, working on projects, interviewing, conducting surveys, organizing and synthesizing their data. Students also present their projects and defend their results. Using pencil and paper, interviews and observation, students may be evaluated on their knowledge of fundamentals of statistics, their ability to plan and carry out projects, and making decisions based on the data gathered. Use of performance tasks may be quite appropriate as an assessment strategy.

M. Resources and references

National Council of Teachers of Mathematics (1989). *Curriculum and Evaluation Standards for School Mathematics*. Reston VA: NCTM.

Sahai, H., & Reesal, M. R. (1992). Teaching elementary probability and statistics: some applications in epidemiology. *School Science and Mathematics, 92*(3), 146-149.

15
PROBABILITY

A. Overview

Concepts included in this chapter are: probability, sample, population, difference between probability and statistics, numerical expression, theoretical probability, and calculating the probability of an event.

B. Background information for the teacher

Most of the topics in school mathematics are the study of "what is" such as $5 + 2 = 7$. Probability is the study of "what might be." Consider how much of our lives are dependent upon uncertainties. To understand anything as basic as a weather report, a student must understand probability. The whole insurance industry is based on an understanding of probability. A mathematician, called an actuary, calculates the likelihood of the occurrence of an event for which insurance is sought, and you pay premiums based on this probability.

Probability and sampling

Probability is the study of how likely it is that something will happen. It is the use of numbers and mathematical reasoning to predict the likelihood of an event in a sample using information taken from a known population.

Samples are items, objects, or pieces of information drawn from a larger body of data. A sample is a subset of the *population*, which is the entire set of all of the possibilities or data points. For instance, if there are 100 jurors in a room, they are the population. If you interview 12 of them, the 12 are the sample. For statistics, it is critical to obtain a representative sample, one that is likely to fairly represent the entire population. Techniques such as random sampling are used to obtain a representative sample.

There is a difference between probability and statistics. Probability reasons from the population to the sample, while statistics generalizes from the sample to the population. Statistics assumes the sample is known, while probability assumes the population is known.

How do we express probability?

We express a probability as a value between and including 0 and 1, as symbolized:

$$0 \leq probability \leq 1$$

If an event <u>will not</u> occur, such as pigs flying on a windless day, the probability is assigned a value of 0. If an event <u>will always occur</u>, such as the sun rising from the east on Earth, its assigned probability is 1. A probability of 0.5 means that the event is likely to happen only 1/2 (50%) of the time; for instance, the likelihood of tossing a head with a fair coin is 1/2.

What is theoretical probability and how do you calculate it?

The *theoretical probability* of an event involves considering all of the possible outcomes of a population and expressing the probability in these terms. To calculate a simple (or *a priori*) probability, begin by figuring all of the possible outcomes of an event. Then count the number of outcomes that represent success. Then divide the number of possible successful outcomes by the total number of possible outcomes. For instance, the theoretical probability of choosing the ace of spades from a standard, 52-card deck of playing cards is 1/52 since there is only one ace of spades. The number 1 represents the event happening out of 52 different possible outcomes. Similarly, the probability of pulling a jack out of the deck would be 4/52, since there are 4 jacks in a 52-card deck. This could be simplified to 1/13, if desired.

C.   NCTM position

K-4

Students will experience problems that explore the concepts of chance, including the concept of "fair" games.

Students will be encouraged to discuss likely events, certain events, and perceptions of "luck."

5-8

Students will devise and carry out experiments or simulations to determine probabilities related to modeling real-world situations.

Students will make predictions that are based on experimental or theoretical probabilities, and appreciate the power of a probability model by comparing the results of the experimental to the prediction of the theoretical probabilities.

Opportunities will be provided for students to understand the pervasive use of probability in the real world.

9-12

Students will use experimental or theoretical probability to represent and solve problems that involve uncertainty.

Simulations will be used to estimate probabilities.

Students will develop understanding of the concept of a random variable.

All students will be able to create and interpret discrete probability distributions.

Students will describe a normal curve and use its properties to answer questions about data sets that are assumed to be normally distributed.

College-intending students will apply the concept of a random variable so they can generate and interpret binomial, uniform, normal, and Chi-square probability distributions.

D. Integrating problem solving, reasoning, communicating, and making connections

Probability is a perfect environment for using a problem solving format. Without real-life problems the teaching of probability can be too abstract. Reasoning with probability is important because many students (and adults) have misconceptions about the likelihood of events based on their feelings.

The problem-solving activities here require reasoning and decision making using probability. Probability is connected with understanding and communicating such real-life situations as weather prediction, insurance, sports, politics, medical evaluations, and games, as well as situations that use statistics. Inheritance of specific genes is based on probability, a fundamental concept in improving crops, ornamental plants, and domestic animals, as well as predicting inheritance of blood types, certain physical characteristics, and genetically related diseases. Problems of population, pesticide resistance, disease, and pollution require an understanding of statistics and probability.

E. Prerequisite skills and knowledge

Students should understand the concepts of area and ratio and should also understand how to collect data from an experiment. The interpretation of this data is a beginning point for using probability.

F. Students' difficulties, confusion, and misconceptions

Shaughnessy (1981) noted that in a coin toss experiment involving three coin tosses people expect that the probability of getting three tails is 1 in 4. This misconception is based on the outcomes, which seem to be: 3H, 2H & 1T, 1H & 2T, and 3T. This erroneous expectation fails to consider that the possibilities shown in the box below.

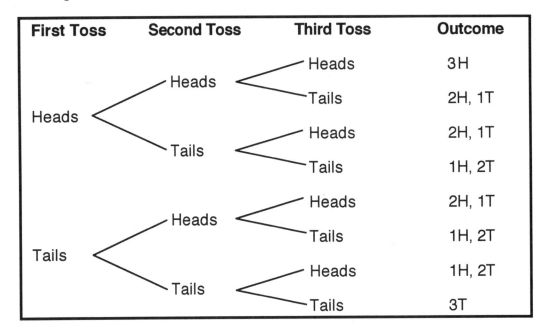

First Toss	Second Toss	Third Toss	Outcome
Heads	Heads	Heads	3H
		Tails	2H, 1T
	Tails	Heads	2H, 1T
		Tails	1H, 2T
Tails	Heads	Heads	2H, 1T
		Tails	1H, 2T
	Tails	Heads	1H, 2T
		Tails	3T

The probability of 3 tails (3T) is 1 in 8 (1/8) while the probability of 2 heads and 1 tail (2H, 1T) is 3 in 8 (3/8). The outcomes listed demonstrate the misconception that even short runs of a coin toss experiment affect the theoretical 50:50 ratio of heads to tails for each <u>outcome</u>, not each <u>toss</u> of the coin.

When a child tosses a coin and repeatedly gets heads she will often say that it is more likely a tail will turn up on the next toss to "even it out." One reason for this is that the child wants their tosses to be representative of the theoretical probability.

Many students disregard the population size. Probabilities drawn from many small samples do not approach the theoretical probability.

Even students with a background in statistics and probability tend to believe outcomes they can think of (their intuitions) will be more likely to occur (Shaughnessy, 1981). For example, if a person forms committees from a group of 10 people who can serve on more than one committee, many students believe that there are more possible committees of 2 people than of 8 people because the committees of 2 are easier to construct. However, the eight non-committee members left over from each 2-person committee construction are the complement of these committees. So there is a one-to-one correspondence between groups of eight and groups of two.

G. Factors contributing to students' difficulties, confusion, and misconceptions

Most people have a poor intuition for probability, often including mathematicians. Calculating probabilities is easy, but most people fail to do it.

Students want their results to match the theoretical probability because many texts have the theoretical probability listed as "the answer" and students want to get correct answers.

It is very time-consuming to run large-scale experiments in classrooms. Many small samples do not approach the theoretical probability, so even when students combine results there may be a misunderstanding of the effect of population size in determining probability.

Many students rely on intuition and do not understand the importance of critically reasoning out the problem.

Since probability is often perceived as "luck," many students do not use what they know and instead rely on what they feel should be true.

H. Appropriate teaching strategies

Appropriate strategies include teaching for conceptual change, learning cycle, discussion and demonstration, and discrepant events. Using the learning cycle in activities such as "Tossing two-color counters" and "Fair Spinners" encourages students to predict, experiment, and evaluate. Discussion and demonstration encourages students to share ideas about probability in "What colors?" and the conceptual change strategy is appropriate for "Assessing real-life situations." "Monty's dilemma" involves discrepant events, where students find their predictions do not fit the outcomes.

I. Teaching notes

"What colors?" is designed for young children who are ready to explore probability for the first time. The goal is to bring about in young children a less intuitive sense of probability. The activity can be repeated with a variety of colors, and with more cubes in the sack with older or more experienced students. The goal is to get students thinking, not to figure out all of the concepts in probability. Teach by asking questions, not providing the answers.

"Tossing two-color counters" is an activity designed for younger students. Two-color counters are available from many different sources, or coins or lima beans colored on one side can be used.

"Fair spinners" is another way to get at the fractional ideas embedded in probability. Familiarity with dividing circles is important in doing this activity.

"Monty's dilemma" is designed for middle and high school students to explore probability. Students might originally have the idea that the odds of winning are 1 in 2 because one of the gag doors was opened; but if no action to change doors is taken, the probability for the door chosen is still 1 in 3. Therefore, the probability for the switch is 2/3. If students still don't see it after they conduct their own experiment, have them spin a three equal-area color spinner and try 100 tries of switch and 100 tries of stay. For fun, and more understanding, you might have students flip a coin to see if they should stay or switch. The coin flip in conjunction with the spinner should yield similar results.

"Assessing the real-life probability" is designed for high school students as an example of ways that probability is employed in real-life problems. The plant example is straightforward and understandable to the student who understands concepts of sexual reproduction and life cycle, including meiosis and gametic combination. The blood typing activity is included for those who enjoy a challenge and have a basic understanding of genetics and blood groups.

J. Materials needed

blank spinners (plastic spinners inside which you can insert a paper to create your own spinner)
protractors (to help create fair-spinner angles)
coins or two-color counters
dice
graph or grid paper
colored cubes
brown paper sack

K. Activities

See the following pages.

1. What colors?

This activity is designed to be a discussion and demonstration. The questions are suggested to encourage conversations about probability.

The teacher should have a sack of five cubes representing three different colors.

Tell the children that there are five cubes in the bag and there are three different colors. Pull a single cube out of the bag and observe the color. Put the colored cube back into the bag. Record the color so the students can keep track with you.

Ask the students what color they predict you will pull out next. If students list a color other than the one shown, ask why they think that color is in the sack. Pull out a cube, record the color, and put it back into the bag.

Repeat the sequence.

Reinforce the concepts of probability by asking questions such as those below.

How many times would I need to pull out a cube to figure out what three colors are present?

If I put the cube back each time, how many times would I have to pull out a cube to know how many of each color is in the sack?

How many do you predict of each color in the sack?

Does your favorite color have to be in the sack?

If you pull out a red cube and put it back, does that mean the next one pulled out shouldn't be a red?

Would I be able to tell what colors were in the sack, and how many there are of each color, if I didn't put the cube back each time? How many cubes would I have to pull out then?

Is it just luck when I pull out a certain color?

2. Tossing two-color counters

What do you think will happen if you toss a single red-green counter 10 times?

> How many times would you expect the red side to be showing?

> Why?

Try the experiment 10 times and record your results.

How many times was the red side showing?

Is this what you expected?

Explain the results in relation to your predictions.

If you toss the counter 40 more times, what do you think will happen?

Explain your prediction.

Toss the counter 40 more times and add these results to the results you found earlier. What are your results for 50 tosses?

Is this what you expected?

Explain the results in relation to your predictions.

If you flip the counter a total of 100 times, what do you predict will happen, based on your results so far?

Explain your prediction.

Record all of the class results for this experiment so that everyone can see them.

In your group, try to make sense of the combined results and compare them to your individual and group results.

Record the results of your discussion.

3. Fair spinners

Imagine you will play the following game using the spinner shown below.

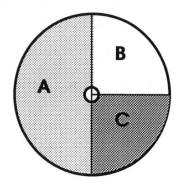

Each of three players chooses a letter (A, B, or C), the order of choice corresponding to the order of the players' birthdays. A player receives a point if the spinner lands on his or her letter during his or her turn.

After ten turns, who do you think will win the game?

Explain your prediction.

Try this game as an experiment to see what your results are.

Were the results what you predicted?

Is the game fair for all three of the players? Explain.

Could this game be made more fair? If so, how?

Use a diagram to show your solution.

Design a fair spinner for four players using the game above. Explain why it is fair.

Create a new game that uses a spinner that is not fair to all of the players.

Have your neighbor explain why your game is not fair while you explain her unfair game.

Create a new fair game that uses a spinner. Explain your game and spinner and why it is fair.

What makes a game fair or unfair?

Is there such a thing as an unfair spinner? Explain.

4. Monty's dilemma[7]

There was a game show on television that had three doors for a contestant to choose among. Doors were numbered 1, 2, and 3. It was set up so that there were two doors with gag gifts and one door with a really big prize behind it.

Imagine that a contestant is asked to choose a door. The game show host, Monty, then opens one of the gag gift doors. Then the contestant is offered the option of either sticking with the original door chosen or switching to the remaining unopened door.

What should the contestant do: stick with the original door chosen or switch to the remaining unopened door?

Explain your reasoning.

If you were the contestant, what would you do?

Why?

In your small group or with a partner, design an experiment to test your advice.

Run it 100 times.

Discuss the results and draw conclusions from your data.

Is it a valid experiment?

Why or why not?

Explain your group's experiment in a class discussion.

Based on your experiment, what would you recommend that the contestant do?

Explain your advice using mathematics.

Did you change your mind from your original advice?

Why or why not?

Explain using mathematics.

---

[7] Idea adapted from Shaughnessy, J. M., and Dick, T, (1991). Monty's dilemma: Should you stick or switch? *Mathematics Teacher, 84*, 252-256.

5. Assessing the probability for real-life situations: the genetics of peas ~ simple dominance, one-trait inheritance [8]

Work collaboratively in small groups or with a partner.

Genetics has many interesting applications of probability. During sexual reproduction, a male gamete (sperm) and a female gamete (egg) fuse, creating a new individual. Each gamete contains one member of each chromosome pair found in the respective parent, so when they fuse, the new individual also has pairs of chromosomes. Since genes are carried on chromosomes, we can apply this pattern to the inheritance of genes, and predict the probability of specific gene combinations in the offspring.

For example, in pea plants, one gene may determine if the dried seeds are round or wrinkled. The *allele* (an alternative form of a specific gene) for round seeds is dominant over the allele for wrinkled seeds; that is, if the round-seed allele (**R**) is present along with the recessive, wrinkled-seed allele (**r**), the seeds will be round. Only if two recessive alleles are present (**rr**) will the seeds be wrinkled.

If the combination **rr** results in wrinkled seeds, what two combinations produce round seeds?

If each gamete of a wrinkle-seeded plant contains the allele **r**, what possible allele can be in each gamete of an **RR** plant? of an **Rr** plant?

Look at the example below:

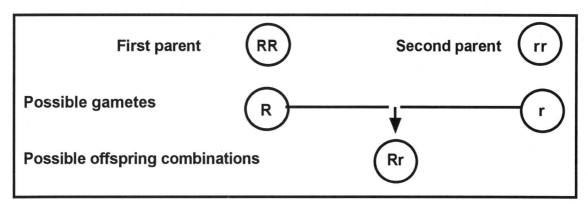

What is the probability that all offspring from this mating will have round seeds? wrinkled seeds? Explain your reasoning.

---

[8]Saigo, B. W. (1998, in preparation). *Conceptual Change Activities in Biology (working title).* Montgomery, AL: Saiwood Publications.

What if the egg of an **Rr** plant combines with the sperm of an **Rr** plant? The problem can be set up as below:

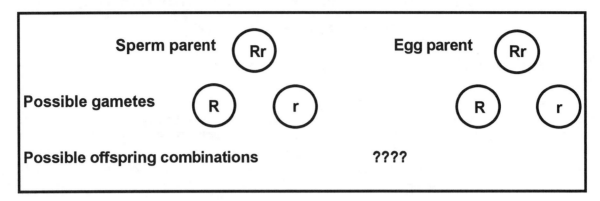

How many different ways do you think the potential gametes from the sperm source could be combined with the potential gametes from the egg source?

Make a diagram to demonstrate your reasoning and show the potential genetic combinations.

Share your ideas in your small group.

Can your reasoning be expressed mathematically?

What proportion of the offspring plants would you expect to produce wrinkled seeds? Explain your prediction.

What proportion of the offspring plants would you expect to produce round seeds? Explain your prediction.

Design an experiment to test your predictions of probability.

What would you do to be sure that the seeds you collected to plant (your sample) were representative of all of the offspring plants (the population)?

What impact would it have on your study if you:

     planted only 8 seeds?

     planted 16 seeds?

     planted 100 seeds?

     planted 1000 seeds?

Do you think it is likely that the data from your study will exactly match your probability predictions for round and wrinkled seeds?

Discuss your experimental design and these questions in your small group, then with the entire class.

6. Assessing the probability for real-life situations: Human ABO blood group [9]

Human blood contains a diversity of proteins, each of which is genetically determined. The first and most common way of identifying blood is by ABO typing, which is of life-and-death significance when it comes to transfusing blood from one person to another.

A person can have blood type A, B, AB, or O. These blood types are four distinct phenotypes resulting from the inheritance of a gene usually designated by the letter I. The convention in genetics is that the dominant form of the gene is represented by a capital letter, and the recessive form is represented by a lower case letter. The I gene has three alleles (alternate forms of the same gene), which are I^A, I^B, and i. The alleles I^A and I^B are codominant, meaning that when they are present together, they are both expressed. The result is blood type AB. The allele i is recessive, meaning that it is not apparent when present with either of the dominant alleles. As a result, both genotypes $I^A I^A$ and $I^A i$ result in a person having blood type A. The genotypes $I^B I^B$ and $I^B i$ both result in a person having blood type B. If two recessive alleles are present, ii, the person will have blood type O.

As a result of meiosis, each gamete carries one gene for ABO blood type. Assuming random processes of gamete formation and subsequent combination of eggs and sperm cells, it is possible to predict inheritance probabilities of blood type. In fact, ABO blood type was commonly used in court cases regarding paternity and criminal identification before the much more precise methods of DNA matching were available.

a.   Parents with the genotypes $I^A i$ and $I^A i$ could have children with the following genes for the ABO blood group[10] :

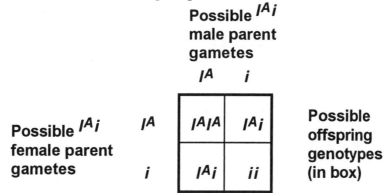

Given the table above, what is the likelihood that a child of these two parents would have blood type A?

---

[9] Saigo, B. W. (1998, in preparation). *Conceptual Change Activities in Biology (working title)*. Montgomery, AL: Saiwood Publications.
[10] Barring aberrations during meiosis.

Explain your reasoning.

What is the probability of any child conceived having blood type O?

What is the likelihood of any child conceived having blood type B?

Explain.

If the parents have four children, will the children necessarily represent all of the blood types in the table?

Explain your reasoning.

Based on your reasoning above, could this couple have four children that all have blood type O?

Explain.

b.   Complete the probability table (also called a Punnet Square) started below for parents having genotypes $I^A I^B$ and $I^A i$.

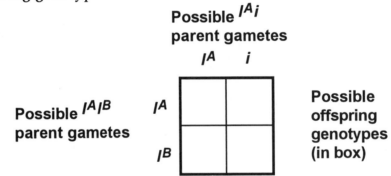

What if these parents have only one child?

What is the probability that the child will be of blood group A?

What is the probability that the child will be of blood group B?

What is the probability that the child will be of blood group AB?

What is the probability that the child will be of blood group O?

c.   Apply the information on the following chart and what you have learned about human ABO blood type inheritance to analyze the situations below.

If a person has blood type (phenotype)	A	B	AB	O
The pair of alleles (genotype) could be	$I^A I^A$ or $I^A i$	$I^B I^B$ or $I^B i$	$I^A I^B$	ii

How many different genetic combinations are there for the ABO blood group?

Create a chart to predict <u>all the possible gamete pairings</u> in regard to blood type. (*Hint:* Since the three possible gamete genotypes are I^A, I^B, and i, your probability table will have three rows and three columns, as shown below.)

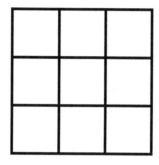

Tally the frequency of each blood type in your chart.

What is the theoretical probability for each blood type occurring?

Convert it to a percentage.

In the U.S. population, would you expect to find these same probabilities?

Explain your reasoning.

Local populations in some Scandinavian countries have a very high percentage of blood type B. How can you explain this?

How might the historical geographic origin of subsets of the U.S. population be a factor in the actual versus theoretical probabilities for blood type distribution in the U.S.?

d. Pretend you are a forensic scientist, applying what you know about human ABO blood groups.

Here is the evidence you have:
A home was broken into, a glass-fronted china cabinet was smashed, and silver candlesticks were stolen. Blood was found on a piece of the glass at the crime scene. It was type A. Suspect X was captured in possession of the candlesticks, and it was discovered that the suspect had type A blood.

Compared to a random sample of 100 persons, what is the probability that the blood drops found at the crime scene could belong to Suspect X?

Do you think knowing the ABO blood type would be enough evidence to prove guilt?

Explain your reasoning and get ideas from others in your small group.

What other information would be useful to <u>support</u> or <u>rule out</u> this evidence?

Share your group's ideas with those of others through a whole-class discussion.

e. Invent other forensic scenarios using inheritance of ABO blood types and exchange them with other groups to solve.

f. Topics related to ABO blood types may be researched with books, articles, the Internet, the American Red Cross, and other sources.

L. Assessment ideas

Student interviews on their conceptions of the odds of "winning" at a game of chance would provide pre- and post-assessment to see if understanding of probability improved.   Probability lends itself to a diversity of performance assessments. Students could create situations using cubes in sacks that would result in given probabilities.   Students could conduct experiments, then describe and explain their results.   Understanding could be developed and assessed by having students analyze a probability-based, real-life problem as reported in the media and also through journal writing about everyday uses of probability.

M. Resources and references

Bennett, J. O., Briggs, W. L., and Morrow, C. A. (1996).   *Quantitative Reasoning: Mathematics for Citizens in the 21st Century.* Reading, MA: Addison-Wesley.

Konold, C. (1996).   Representing probabilities with pipe diagrams. *Mathematics Teacher, 89,* 378-382.

Mendenhall, William (1987). *Introduction to Probability and Statistics.* Boston, MA: PWS.

National Council of Teachers of Mathematics. (1989). *Curriculum and Evaluation Standards for School Mathematics.* Reston VA: Author.

Saigo, Barbara W. (1998, in preparation). *Conceptual Change Activities in Biology* (working title). Montgomery, AL: Saiwood Publications.

Schulte, A. and Smart, J. R. (Eds.) (1981). *Teaching Statistics and Probability : Yearbook of the National Council of Teachers of Mathematics.* Reston, VA: NCTM.

Shaughnessy, J. M. (1981). Misconceptions of probability: From systematic errors to systematic experiments and decisions. In A. Schulte and J. R. Smart (Eds.), *Teaching Statistics and Probability*: 1981 Yearbook *of the National Council of Teachers of Mathematics.* Reston, VA: NCTM.

Shaughnessy, J. M., & Dick, T. (1991). Monty's dilemma: Should you stick or switch? *Mathematics Teacher, 84,* 252-256.

Shaughnessy, J. M. (1992). Research improbability and statistics: Reflections and directions. In D. A. Grouws (Ed.) *Handbook of Research on Mathematics Teaching and Learning, 465-498.* New York: MacMillan.

16
TRIGONOMETRY

A. Overview

Concepts included in this chapter are: trigonometry, sine, cosine, tangent, period, amplitude, wavelength, frequency, ways to represent trigonometric ideas, and applications of trigonometry.

B. Background for the teacher

Trigonometry is a branch of mathematics which originally meant triangle measurement. Now it includes features of the circle and wave mechanics. The history of trigonometry dates back to around 100 B.C. Two Greek astronomers, Hipparcus of Nicaea and Ptolemy, were the first to make important advances in trigonometry.

The mathematics and science of music, earthquakes, springs, revolution of planets, carpentry, and aeronautics are only a few examples of how we rely heavily on trigonometry. Any problem which uses the right triangle or studies waves or periodic motion makes use of trigonometry. Seismologists, musicians, physicists, artists, electronic engineers, acoustical engineers, television specialists, architects, and structural engineers are among those who use trigonometry in their professions.

The language of trigonometry

Terms such as sine, cosine, tangent, period, amplitude, wavelength, and frequency, are associated with trigonometry. They express the relationships between angles and sides of a triangle.

For our purposes here, we will use a familiar right triangle having sides of 3, 4, and 5, and will define trigonometric functions in terms of acute angle x in this triangle. The *sine* of x (sin x) is the ratio of the side <u>opposite</u> angle x to the hypotenuse. This ratio is a number. For our triangle, therefore, sin $x = 3/5$.

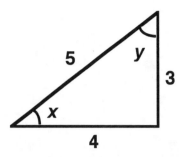

The *cosine* of x (cos x) is the ratio of the side <u>adjacent</u> to angle x to the hypotenuse. For our triangle this will be 4/5, so the equation would be cos $x = 4/5$.

The *tangent* of x (tan x) is the ratio of the side <u>opposite</u> to the side <u>adjacent</u> to angle x; *i.e.*, tan $x = 3/4$.

Using these same definitions, sin $y = 4/5$, cos $y = 3/5$, and tan $y = 4/3$.

If we know the ratios (*i.e.*, sin or cos of an angle) we can use a table or calculator to determine the angle itself.

The Laws of Sines and of Cosines

The *laws of sines and cosines* give the relationships between the sides and angles of <u>any</u> plane triangle. Use the triangle below as a reference.

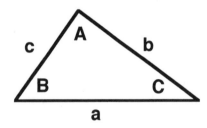

Law of sines:
$$\frac{a}{\sin A} = \frac{b}{\sin B} = \frac{c}{\sin C}$$

Law of cosines:
$$a^2 = b^2 + c^2 - 2bc\cos A$$

If $\hat{A}$ is 90°, cos 90 = 0. As a result, we get the familiar Pythagorean relation, $a^2 = b^2 + c^2$. Using the Pythagorean relation, we can also show that $\sin^2 2A + \cos^2 2A = 1$.

We can use a *unit circle* as well. The unit circle is a figure that is used to introduce radians. One revolution around the circle gives us two related measures, an angle of 360° and the length of the arc equal to the circumference of the circle.

The circumference of a circle is $2\pi r$ (r being the radius of the circle). If r = 1 unit (called unit circle), the circumference (length of the arc for one revolution) is 2π in radians. This arc distance corresponds to an angle of 360°. Since $\pi = 3.14$, 360° corresponds to 2π or 6.28 radians. Therefore, one radian is $\cong 57.3°$.

The *circumference* of the unit circle is **2π radians.**

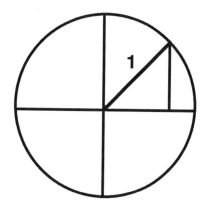

Any repeated (periodic) motion, such as a pendulum, electrical current, sound, ocean waves, and the vibrations of an earthquake, can be presented pictorially. An oscilloscope helps us see the pattern of a periodic motion.

A swinging pendulum follows a repeated motion. The pictorial representation of any periodic motion can be represented as a trigonometric curve (representation). Following the motion of an object around the circle and plotting the angle it makes with the horizontal against the value of trigonometric function would yield the familiar pattern seen on an oscilloscope.

The periodic motion of an object or a signal around a circle may be represented as a plot of the angle it makes with the horizontal and the value of the trigonometric function. The circumference of a circle of radius r is $2\pi r$. The distance around a *unit circle* ($r = 1$) is 2π. This motion around the circle corresponds to 360°.

If we plot the value of the angle that the object makes with the horizontal axis in multiples of π (2π is a complete revolution in radians), π is half the revolution, and $\pi/2$ is one-fourth of the revolution; therefore, 2π radius corresponds to 360°; π is 180°, and $\pi/2$ is 90°.

The values of sine for these angles are:
sin 0 = 0, sin $\pi/2$ = 1, sin π = 0, sin $3\pi/2$ = -1, sin 2π = 0.

Plotting the value of the angle against the value of sine function yields the following graph, called the *sine graph*.

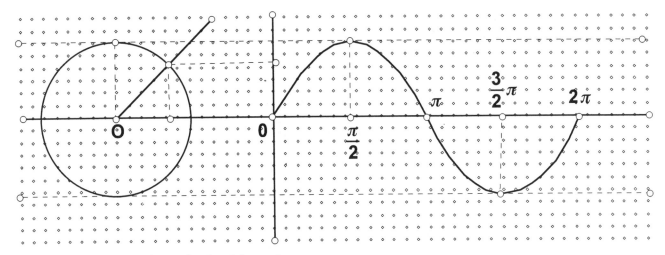

A wave can be described with various terms.

The *amplitude* of a sine curve is 1/2 the vertical distance between a crest and trough.

The *wavelength* is the distance between two adjacent crests or troughs.

Frequency is the number of complete waves passing a fixed point along the wave each second.

The *period* is the time it takes for one wavelength to pass by.

C. NCTM position

9-12

Students will apply trigonometry to problem situations involving triangles.

Students will explore periodic real-world phenomena using trigonometric functions.

D. Integrating problem solving, reasoning, communicating, and making connections

As students go through this chapter on trigonometry, they encounter real problems. They are challenged to plan, think of appropriate strategies, search for solutions, and communicate findings. The activities in this chapter are designed to help students to connect the principles of trigonometry to situations like astronomy, physics, carpentry, surveying, farming, and sports.

E.   Prerequisite skills and knowledge

The activities and exercises in this chapter build on students' knowledge of arithmetic, simple geometry, and the language of algebra.

F.   Students' difficulties, confusion, and misconceptions

It is difficult for many students to accept that trigonometric functions such as sine, cosine, and tangent are ratios that express relationship and do not have units associated with them.

For many students, it is difficult to accept that an angle formed by two sides is independent of the length of the sides.

Many students have difficulty visualizing trigonometric functions.

G.   Sources of students' difficulties, confusion, and misconceptions

Classroom and textbook presentations assume that just because the formulas have been given, the students will immediately understand and accept them.

Our heavy reliance on proofs, definitions, and rules does not allow the students to experience the real meaning of the concepts.

H.   Appropriate teaching strategies

Teaching for conceptual change, discussion, and mental model building are important strategies for developing an understanding of trigonometry. We recommend using the conceptual change strategy for activities 1, 2, and 10. Collaborative learning may effectively be used with activities 3, 4, 5, 6, 7, 8, and 11.

I.   Teaching notes

Waves of all kinds, including waves in strings, water waves, sound waves, and electromagnetic waves, among others, are expressed in terms of trigonometric functions. These are important topics; however, they are beyond the scope of this book. In this chapter we will concentrate primarily on the trigonometry of the right triangle.

J. Materials

protractors
rulers
measuring tapes
2 x 4 boards of various lengths
spring scales or other force measurers
compasses (for drawing circles)
graph paper

K. Activities

See the following pages.

1. How tall is the flag pole?

How can you use trigonometry to measure the height of the school flag pole?

Share your plans in your group and present them to others in a class discussion.

Implement your plan and share your findings.

2. Shooting a projectile

What if you were to shoot an object through the air with a velocity of 5m/sec at an angle of 30 degrees from the horizontal?

In your group, come up with a plan to tackle the following questions:

> How high will the object rise?
> How far will the object travel?
> How long will it take the object to hit the ground?

Share your ideas with others.

How do your ideas compare with other groups in class?

Work on answering these questions, then compare your work with that of others.

3. So, you are a carpenter!

You want to put a shed in the back yard. You must precut the rafter ends so they will be vertical when in place. The front wall of the shed is 12′ high, the back wall is 9′ high, and the distance between the walls is 12′. At what angle should you cut the ends of the rafters?

In your group, address the following questions:
> What is given?
> What is to be done (found)?
> What is your plan to approach the problem?
> What operations do we use to solve the problem?

Present your group's plan to the rest of the class.

Carry out the plan.

Compare your results with the results of other groups. Discuss the various strategies that were used.

4. Exploring the skies

A laser beam is directed toward the Moon. It misses the center by 1 degree. How many miles will it be from its assigned target?

In your group, respond to the following questions:

What is the question (problem) we are investigating?

What is given?

What is to be found?

Do we have all the information? If not, what is missing?

How can you get the missing information?

What is your plan?

What operations should be used?

Carry out your plan.

Compare your results with the results of other groups.

Discuss the various strategies that were used.

5. Congratulations on your job as a surveyor

Your group is on a mapping expedition and needs to find the height of a hill.

Decide on a plan to find the height. (*Hint:* One way to do this is by making two observations, from points A and B).

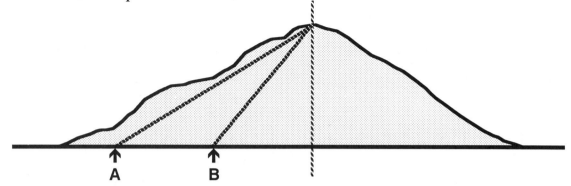

Do you have or can you find all the information you need? If not, what do you need and where can you find the information?

What materials do you need to carry out your plan?

Share your plan with others in the class.

Get the necessary materials and find the height of the hill.

Compare your results with others in class.

Discuss the various strategies that were used.

If you were to do the project over, what would you do differently?

6. Where is it?

Using a map of the school, protractor, compass, and meter stick, how can you use trigonometry to locate a particular place or item in the school <u>chosen by your teacher</u>?

In your group, develop a plan.

Share your plan with the class.

Get the necessary materials and implement your plan.

Share your results with others in the class.

What made your search successful or unsuccessful?

7. So, in the final analysis, which way will it go?

Imagine a box with forces acting on it, as shown in the diagram. The length of the vector represents its magnitude. For example, vector a ($\vec{a}$) has a magnitude of 15 units.

Which way will the box move (magnitude and direction) in response to the forces?

In your group, discuss a plan as to how you will proceed to solve this problem.

Do you have all the information you need? If not, where can you get the information?

Implement the plan and share your findings (numbers and drawings) with others in class.

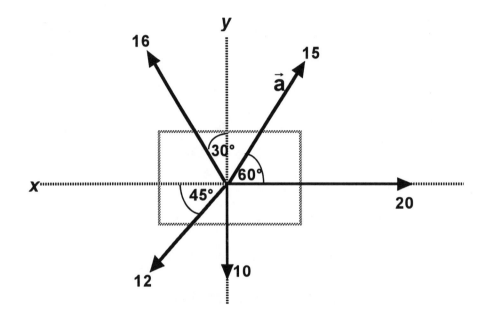

8. Measuring the radius of Earth

Within your group, come up with a plan to measure the radius of Earth. Use such information as the height of a mountain given by your teacher or found in a book and the angle formed by the top of the mountain with the vertical.

Share your ideas with others.

Get the necessary material and data and make the calculations.

What results would you have obtained if you had chosen a different mountain?

Explain.

9. You just got a telescope

Assume that you must incline your telescope 20 seconds of an arc to see a star directly overhead. Your telescope moves with the Earth, while the star is fixed.

Using this information, how can you find out how fast light travels from the star to you?

In your small group, discuss the following questions:

> What is the question or problem you are investigating?
>
> What do we want to find?
>
> What is given?
>
> Do we have all the information needed to make the calculation?
>
> If not, what other information is needed and where can we find it?

In your group, design a plan to tackle the challenge.

Share your plan with others in the class.

Implement the plan.

How would you check your results?

10. Find the width of a river

You have been challenged to find the width of a river. You cannot cross the river.

What measurements do you need to make?

What materials do you need to make the measurements?

Design and carry out a plan to find the width of the river.

Share your findings with others.

11. Law of cosine

Your challenge is to measure the sides and the angles of a triangle and check the results by using calculations.

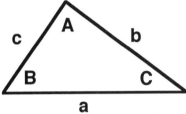

In your group, plan on what materials you need.

Carry out the appropriate measurements, then check your data with others.

Decide on the formulas you will use to make the calculations. Use the measurements from two sides and the angle between them to calculate the length of the third side.

How did your measurements compare with your calculations?

If there were discrepancies, what may be causing this?

As a group decide how you could find the angles of a triangle if you knew the sides.

Share your ideas with others.

How would you test your plan?

12. Your project

This is your turn to come up with your own project. You are to use trigonometry in solving the problem you have posed. Choose a project in which you are really interested. Carry out the project and make a presentation to the class.

L. Assessment ideas

Pencil-and-paper assessment may be used to assess students' knowledge of trigonometric functions and the laws of sine and cosine. These assessments may include problems and situations where students are asked to calculate surface area and volume of different 3-D shapes, provide additional examples of 3-D shapes in their daily lives, or write about the importance of the concept. Also, pencil and paper assessment may be used to assess students on applying trigonometry to everyday situations.

Students' performance on their activities and projects may be good indications of their conceptual understanding of basic concepts of trigonometry.

M. Resources and references

Bonsangue, M. (1993). A geometrical approach to the six trigonometric functions. *Mathematics Teacher, 86*(6), 496-498.

Boyes, G. R. (1994). Trigonometry for non-trigonometry students. *Mathematics Teacher, 87*(5), 372-375.

DiDomenico, A. S. (1992). A trigonometric exploration. *Mathematics Teacher, 85*(7), 582-583.

Lewis, G. (1994). The lost trigonometry class and the hidden treasure. *Mathematics Teacher, 87*(1), 19-22.

National Council of Teachers of Mathematics. (1980). *A Source Book of Applications of School Mathematics*. Reston, VA: NCTM.

National Council of Teachers of Mathematics. (1989). *Curriculum and Evaluation Standards for School Mathematics*. Reston VA: NCTM.